P9-BXW-343

A YEAR LIKE NO OTHER

A YEAR LIKE NO OTHER

HOW A GLOBAL PANDEMIC LED TO VANDERBILT UNIVERSITY'S PROUDEST MOMENT

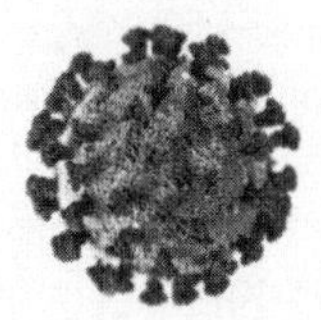

RYAN UNDERWOOD

Foreword by CHANCELLOR DANIEL DIERMEIER

A Year Like No Other
How a Global Pandemic Led to Vanderbilt University's Proudest Moment

Unless otherwise noted, material for this book came from interviews conducted between December 2020 and February 2021, as well as from minutes taken at Vanderbilt Board of Trust meetings held throughout 2020.

Published by Forefront Books.

Cover Design by Bruce Gore, Gore Studio Inc.
Interior Design by Bill Kersey, KerseyGraphics

ISBN: 978-1-63763-003-7 print
ISBN: 978-1-63763-004-4 e-book

CONTENTS

FOREWORD

In July 2020, when I officially began my role as Vanderbilt's ninth chancellor, we were in the eye of a perfect storm.

Three crises had been stirring in unison: that of public health, as the COVID-19 virus unrelentingly took lives and upended livelihoods around the world; the resulting financial downturn, which over the year would lead to the loss of more than 650,000 jobs in higher education alone; and the moral crisis in our nation, evident in the bitter polarization of American politics and the renewed urgency to fight for racial justice. Each crisis alone would have required all our efforts. Daunting, important, and urgent, we needed to address them all at once.

Between February and April 2020, Vanderbilt's endowment dropped nearly 20 percent, a significant loss at any time, but felt more acutely due to the growing expenses for technology and personal protective equipment, accompanied by lost revenue across the board.

Concurrently, students had left our campus in a rush. Graduate students had to immediately adapt to remote strategies to complete their studies and dissertation projects. Postdoctoral fellows swiftly reenvisioned their approaches where possible. Student-athletes lived with uncertainty about when and how their next competition would occur—and about what a prolonged break could mean for their success on and off the field.

Life-changing decisions had to be made, and in rapid succession, but little clarity existed—even in regard to the disease itself. The CDC's initial hypothesis that masks did not stop the spread of COVID-19 was revised as evidence evolved, and masks were deemed a critical, often mandatory, precaution. The members of Vanderbilt's Plant Operations team, as well as faculty and academic leaders, were losing sleep over the possibility that the virus could travel through heating and air conditioning systems, which would then make classrooms irrepressible hotbeds for infection. Thankfully, this particular nightmare did not come to pass.

Underlying the profound uncertainty was yet another loss: the gradual disintegration of the rhythm of university life. From the moment Vanderbilt's in-person commencement was rescheduled for the class of 2020, the familiar pacing of milestones was gone—the excitement of move-in day in August, fall football games with packed stadiums, the final academic push before the holidays, the buildup to spring break, and graduation in May.

The operational and financial challenges were daunting. Our budgets, the result of months-long processes, had to be redone in mere weeks. The logistics around storing thousands of students' personal belongings—from clothing to bicycles—became an uncharted priority. The admissions process was suddenly subject to new calendars and considerations, including the cancellation of all in-person campus visits and the rapid-fire development of a personalized virtual tour. As buildings were locked up, communications increased to relay new expectations and timelines.

The summer months, when faculty would usually be able to focus on their research, were suddenly crammed full of virtual town halls about hybrid teaching, online resources, and classroom protocols.

This disrupted pattern—and an overall landscape of immense uncertainty—was what I stepped into on July 1, 2020. It was an unusual first day for many reasons, but the work had started earlier.

Under normal circumstances, the late spring and early summer would have included a gradual wind-down from my previous role as provost at University of Chicago, paired with information-gathering phases and meetings with key Vanderbilt stakeholders. Instead, the transitional period was marked with prolonged meetings in the university's virtual war room, where I witnessed the incredible work being done by Interim Chancellor and Provost Susan R. Wente, Vice Chancellor Eric Kopstain, and other leaders to address some of the most monumental questions that Vanderbilt had ever faced.

It was not lost on my friends and former colleagues that I had taken the helm at Vanderbilt at an alarmingly unprecedented time. Indeed, I received messages of concern—they wondered how I was faring, if I missed Chicago, if I wished the timing had been different.

My answer to them was, and remains the same to this day: if one is privileged enough to lead such a distinguished institution as Vanderbilt, it's even better to do so when the stakes are high. Better to be prime minister of Britain from 1940 to 1945 than from 1950 to 1955.

Crises can bring out the best in us, and they leave lasting memories. It is during these moments that humans often reach their full potential and transform the most profound challenges into their proudest moment.

In that season at Vanderbilt, little was known and little was certain. So we looked to our mission to guide us. We knew we would do whatever we could to advance the interdisciplinary, pathbreaking research that was now more important than ever. And we would find a way to foster the residential education that brought students together and enabled them to learn from peers with different perspectives, to challenge their points of view, and to experience firsthand what it means to do what is needed—even if that path defies simplicity or easy answers.

This very mission was at the forefront of Founders Walk, a rite of passage for first-year students at the start of the fall semester. In August 2020, for the first time in Vanderbilt's history, students engaged in the event virtually—whether from their dorm rooms or their homes away from campus. Speaking to a video camera, I told them that our decision to keep the doors to our campus open and to invite students and faculty back for in-person learning was difficult. The path ahead would not be easy.

In fact, throughout the lead-up to the fall semester, I was often reminded of John F. Kennedy's iconic speech from 1962, made just after his visit to NASA's newly opened space center in Houston. To a crowd of more than thirty thousand people at Rice University, including many students, he said, "we choose to go to the moon . . . not because [it is] easy, but because [it is] hard. Because the goal will serve to organize and measure the best of our energies and skills, because that challenge is one that we are willing to accept, one that we are unwilling to postpone, and one which we intend to win."[1]

Vanderbilt, too, chose the hard path. We were tested—as individuals and as a university community. We faced obstacles and temporary setbacks, clusters of positive test results, and moments of reckoning. No matter how history ultimately judges our actions, it is important that we capture this period for posterity.

Committed to a common purpose, we pushed ourselves more expansively than ever before. Staff members often worked seven days a week, sacrificing their personal lives for our shared mission and well-being. Faculty adapted to new platforms, while their impact—as teachers and mentors—became more important than ever before. Students showed resilience despite new, unknown limitations on student life that threatened to derail a pivotal and precious time of personal and intellectual growth.

Without all of these people—our people—we would not be where we are, ready and eager to share this story. I will be forever grateful for what we accomplished together.

—Daniel Diermeier,
Vanderbilt Chancellor

PART 1

THE STORM BREAKS

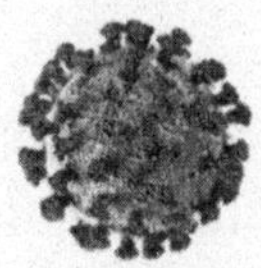

Susan R. Wente first learned of the virus in December 2019—not in her role as the head of Vanderbilt University, but as a scientist.

Online reports of a "novel coronavirus" causing dozens of cases of pneumonia in Wuhan, a city of eleven million in central China, were circulating among cell biologists like Wente, who, after shifting from a distinguished career in science to join the upper ranks of Vanderbilt's leadership, continued to run an internationally respected research program on campus. Her lab had made important discoveries regarding how RNA molecules travel within cells and how different viruses alter those pathways during infection. The pathogen that had emerged in Wuhan was the kind of RNA virus Wente's lab studied, so she read about it with interest.

Assessing the scant data coming out of China as best they could, Wente and most of her peers viewed the virus as something to watch closely but that, with luck, would be contained quickly, like the coronavirus that had caused the global SARS outbreak in the early 2000s.

As a scientist, Wente could observe with curiosity from half a world away as facts about the new virus came to light. But as someone serving double duty as Vanderbilt's longtime provost and its current interim chancellor, she was compelled to consider taking action—even if as just a precaution against what seemed like the remote possibility that the virus could become a danger to Vanderbilt's people and assets.

"The issue," she recalls, "[was] that there just wasn't a lot of information coming out to enable us to determine [the threat], scientifically or institutionally."

Wente continued to track online discussions of the virus as 2020 arrived and, with it, some thirteen thousand students returning to Vanderbilt's campus for the spring semester. Before long, what had begun as a handful of reports from a city many Americans had never heard of built slowly into a steady drumbeat of increasingly concerning news. On January 11, Chinese state media reported that a sixty-one-year-old man had died of an illness caused by the novel coronavirus—the first such death to be reported. Fewer than ten days later, coronavirus cases were reported in Japan, South Korea, and Thailand. The first US case was confirmed near Seattle on January 21.[2] And on January 23, the Chinese government locked down the residents of Wuhan in a belated attempt to keep the coronavirus from spreading.[3]

It was about this time that the team in Vanderbilt's Office of Emergency Preparedness, which had also been monitoring reports of the virus, decided it was time to brief their boss, Vice Chancellor for Administration Eric Kopstain.

"They gave me a deck of slides, and we had a very interesting conversation," Kopstain recalls. "We concluded with two decisions: One, let's keep a close eye on this and hope we don't have to talk a lot more about it because it will be contained. But let's also establish a coronavirus commission with people from across the university."

The commission's nearly forty members would represent virtually the entire university administration—from public safety, finance, and academic affairs to communications, IT, legal, and dining services. The plan was for the panel to convene at the end of the month to begin developing a playbook that university leadership could draw from in the event that the coronavirus came to campus. The team wouldn't be starting from scratch—the broad, earlier planning for an outbreak of any dangerous contagion gave it a running start.

Across West End Avenue from Kopstain's office, at the Loews Vanderbilt office plaza, Vanderbilt Director of Strategic Communications Princine Lewis was also tracking news of the virus. With outbreaks now prominent in the headlines—including possible cases at Tennessee Tech and Texas A&M—parents and others were bound to have questions about Vanderbilt's strategy for protecting students. Already a local television reporter was asking about the implications for the university's study abroad program. The university needed a communications plan, Lewis thought, and fast.

On January 28, the same day Kopstain got his briefing, Lewis went to Associate Vice Chancellor for Strategic Communications Ian Morrison with two recommendations: Vanderbilt should increase its communication about the virus beyond occasional bits of news on the Office of Emergency Preparedness web page. And it should start right away, posting a message to the university's official news site. Lewis and Morrison agreed that the matter of the virus was growing serious enough that its daily handling ought to be escalated to the highest level of Vanderbilt's administration—the ten vice chancellors who oversaw all aspects of the university's operations. Morrison planned to call their boss, Vice Chancellor for Communications Steve Ertel, that night.

When he reached Ertel, it was a difficult conversation—not only because of the subject matter, but because Ertel was at a noisy roller rink with his son's grade-school class and had a hard time hearing. But he heard enough to understand Morrison's concerns and agreed to take them to Kopstain right away. Searching for the quietest place in the rink, Ertel wedged himself into a nook housing an ATM. After speaking with Kopstain, he next dialed Lewis, who had to dash out from the salon, her hair still wet, to take the urgent call from the privacy of her car. Ertel then connected both Lewis and Morrison with Kopstain, who listened as they described their concerns and the near-term communications plan Lewis outlined. He agreed to run with their recommendations and to raise the issue among the other vice chancellors.

"That was the beginning of a more concerted focus among the leadership of the university," Ertel says.

The next day, the first item about the coronavirus appeared on the Vanderbilt news site beneath the headline "University monitoring coronavirus outbreak." "At this time, there are no reports of coronavirus cases on the Vanderbilt campus, and there are no confirmed cases in Tennessee," the report noted. It went on to say that student requests for travel to China for university-sponsored activities would be subject to additional review and that faculty and staff should reconsider nonessential travel to the country.[4]

Despite the growing concern behind the scenes, the new semester was getting off to a normal start. During the week of January 26, in addition to the daily hum and shuffle of classes, lab activity, and office work, there was a screening of a documentary called *A Free Trip to Egypt.* Several PhD candidates defended their dissertations. The Wond'ry, Vanderbilt's Innovation Center, hosted an evening of live music and original poetry. The women's swim team traveled to a meet at the University of Arkansas–Little Rock, men's tennis faced Indiana, the women's basketball team took on Alabama and Tennessee, and the men's hoops squad traveled to Lexington to square off against the University of Kentucky. The Blair School of Music offered several master classes, and the university's InclusAbility organization sponsored a presentation called "Deaf and Awareness

Culture 101."[5] Tickets were moving briskly for upcoming talks with former national security advisers Susan Rice and John Bolton and broadcast journalist and Vanderbilt family scion Anderson Cooper, among others.[6] For most people, as Assistant Vice Chancellor for Plant Operations Mark Petty remembers, the coronavirus was "this thing over there. It was not here."

On January 30, the World Health Organization declared the novel coronavirus to be a global emergency.[7] Three days later, the Philippines reported that a man had died of the virus—the first death reported outside China. The *Diamond Princess*, a cruise ship, was anchored off the coast of Japan, its 3,700 passengers quarantined after an outbreak onboard. More than 700 eventually tested positive, and 14 of them would later die.[8]

On February 11, the virus got its official name: severe acute respiratory syndrome coronavirus-2, or SARS-CoV-2 for short. The illness it caused was named too: COVID-19.[9] France, on Valentine's Day, reported Europe's first COVID-19 death.[10]

For an early February meeting of the Vanderbilt Board of Trust, Wente invited Dr. William Schaffner, a professor of preventive medicine and infectious disease at the Vanderbilt University School of Medicine, to brief trustees on what was known about the virus. That same month, Dawn Turton, an associate provost in Wente's office, called Vanderbilt professor Dr. Mark Denison after Vanderbilt University Medical Center CEO Dr. Jeffrey Balser suggested she invite Denison to speak to the university's coronavirus commission. Denison had studied coronaviruses for more than three decades and was one of the world's foremost experts on the pathogens. When Turton reached him, he was waiting to catch a flight back to Nashville from Washington, DC, where he'd met with infectious disease experts at the US Centers for Disease Control and Prevention and the National Institutes of Health. The next day, Denison stood in front of the commission and told them what he knew. His assessment was sobering, Turton remembers. Denison dispelled any hopes that the outbreak might be seasonal. More distressingly, he explained that the virus could apparently be spread by people who had no symptoms. A person could have the virus and never know it, appearing perfectly healthy. Others could contract it simply by being near enough to that person.

"How would you deal with that in a student population?" Turton recalls wondering. "I didn't see the way forward." Hopes that this virus would go the way of SARS were fading rapidly.

As administrators learned more, Wente stayed in close contact with Bruce Evans, the chair of Vanderbilt's Board of Trust, to keep him informed about the commission's thinking. She also made regular calls to incoming chancellor Daniel Diermeier. In early December, the board had appointed Diermeier, the University of Chicago provost, as the permanent replacement for longtime Vanderbilt chancellor Nicholas Zeppos, who had retired the previous August. Diermeier was not scheduled to assume the chancellor's role until July 1, but Wente kept him apprised of Vanderbilt's decision-making.

Meanwhile, the virus spread around the world, with hotspots emerging in Italy and Iran. By February 27, more than 82,000 SARS-CoV-2 infections had been reported in forty-seven countries, with more than 2,800 reported deaths.[11] In the US, the CDC warned Americans to begin preparing for the virus to spread.[12]

As the strict lockdown in Wuhan continued, and as Italian authorities issued stay-at-home orders for several of that country's hard-hit provinces, the prospect of similar shutdowns in American cities began to look possible. For the nation's colleges and universities, such an unprecedented move could have enormous, even existential consequences. How, in a worst-case scenario, could students be locked down or quickly evacuated from campus? How could professors pivot to online learning? How could research laboratories, America's mighty innovation engine, simply idle? How might a shutdown, and the resulting loss of revenue, affect a university's finances? These were questions no administrator wanted to think about, but in February 2020, they were questions that demanded answers.

At the Forefront of Research

As university leadership began dealing with the realities of the coronavirus heading toward their campus, faculty researchers and their colleagues at Vanderbilt University Medical Center were contributing to the fight against the pandemic at the local, national, and global levels, joining their colleagues worldwide to lend a hand as soon as news of the virus broke. Researchers from all of the university's schools and colleges were involved,[13] in fields ranging from microbiology, engineering, and psychology to anthropology, history, public health, and education.

At the center of some of the most important efforts was Denison. His lab had been funded for more than thirty years by the National Institutes of Health to investigate the replication, pathogenesis, and evolution of coronaviruses—including SARS, MERS, and now SARS-CoV-2. Since he began studying coronaviruses in 1984, Denison's lab had made several seminal discoveries in their biology and in the development of antivirals and vaccines.[14]

Early on, the Denison Lab was a key player in several vital efforts that the entire world was watching with hope. One was the development of a COVID-19 vaccine. Denison and his team were preparing to lead Phase 1 of human testing of mRNA-1273, a vaccine candidate developed by Cambridge, Massachusetts-based biotech company Moderna. This particular vaccine candidate was built on work going back to the 1990s by Dr. Barney Graham, a Vanderbilt PhD alumnus who had been a professor of microbiology and immunology at the university, and who was now deputy director of the National Institutes of Health Vaccine Research Center. Based on a concept devised by Graham, mRNA-1273 was thought to work by triggering the production of antibodies that attack the "spike" protein that enables SARS-CoV-2 to infect cells in the body. By blocking the spike protein, researchers believed, the antibodies "neutralized" the virus.[15]

Vanderbilt's central role in COVID-19 vaccine development was no accident. Researchers like Graham and Denison had built on a century of foundational work, including pathology professor Dr. Ernest Goodpasture's propagation of viruses in chicken eggs in the early 1930s. When COVID-19 emerged, Vanderbilt researchers were ready with knowledge and innovations amassed over decades. "People say, 'How could they make vaccines for corona so fast?'" says Vanderbilt professor and researcher Dr. James Crowe. "Well, they did it based on more than thirty years of research on RSV, HIV, and MERS."[16]

Denison and his colleagues were also involved in investigating the antiviral drug remdesivir, a potential COVID-19 treatment that was in clinical trials in the United States and China. The Denison Lab had researched remdesivir since 2014; Andrea Pruijssers, an assistant professor of pediatrics and the lead antiviral scientist in Denison's lab, had been the first to demonstrate that the drug was effective against SARS-CoV-2.[17] The pharmaceutical company Gilead Sciences would begin testing remdesivir in Phase 3 clinical trials in March 2020.[18]

And the previous November, Pruijssers had reported the first evidence that EIDD-1931, an antiviral developed at Emory University, blocked replication of a broad spectrum of coronaviruses in laboratory tests and prevented the viruses from developing resistance against it. Denison and his colleagues also contributed to a study at the University of North Carolina–Chapel Hill, which showed that a form of EIDD-1931 that could be taken orally prevented severe lung injury in infected mice. Human clinical studies on that therapy were scheduled to begin in the spring.[19]

Meanwhile, at the Vanderbilt Vaccine Center, not far from Denison's lab, Crowe and Associate Professor Robert Carnahan had moved sleeping cots into their lab so they could work around the clock with an international team of academic, governmental, and corporate partners to identify and analyze antibodies that could be used against SARS-CoV-2. Crowe was a world leader in the development of monoclonal antibodies, which target specific viruses; he had already developed them for highly contagious viruses like HIV, dengue, influenza, and Ebola. Crowe's efforts in 2015 and 2016 to rapidly develop a protective antibody treatment for Zika virus had laid the groundwork for his team to move quickly to combat COVID-19.[20]

In addition to the work by Denison, Crowe, Carnahan, and their colleagues, VUMC was leading two clinical trials to determine the safety and efficacy of hydroxychloroquine in treating COVID-19. The drug was well known as a treatment for malaria and rheumatologic conditions, and lab studies had suggested that it might be useful against COVID-19. Some physicians had already turned to hydroxychloroquine in the absence of other therapies. Their doing so sparked some controversy, because the US Food and Drug Administration hadn't approved the drug for treatment of COVID-19. Confusion would increase—and become surprisingly political—when President Donald Trump repeatedly touted hydroxychloroquine's use despite warnings from his own health experts. The two trials VUMC was preparing to conduct in the winter of 2020 were intended to provide sorely needed data on the use in humans of hydroxychloroquine for COVID-19.[21] By the fall, scientists at Vanderbilt and elsewhere had concluded that the treatment was not effective.

Vanderbilt and VUMC faculty were making a difference beyond the pharmacological response to COVID-19. An interdisciplinary team was devising a solution to help increase the supply of scarce ventilators needed for the sickest patients: a fabricated, open-source ventilator design. Led on the university side by the School of Engineering's Assistant Research Professor Kevin Galloway and Mechanical Engineering Professor Robert Webster, the team would test two prototypes by spring, aiming for its ultimate goal of making a design publicly available for anyone to replicate.[22]

At about the same time, Natalie Robbins, a staff researcher with the Vanderbilt Initiative for Interdisciplinary Geospatial Research, was developing an online map of coronavirus infections in Tennessee that was considerably more detailed than others available at the time. In addition, the map provided metrics for each county that were relevant to public health efforts, like the number of families with children, the number of elderly residents, the percentage of uninsured, and more.

Inspired by Robbins's map, anthropology doctoral student Gabriela Oré Menéndez built a similar one for her native Peru. Both maps were shared widely on social media and made available for any public health agency to use.[23]

Meanwhile, Vanderbilt infectious disease experts like William Schaffner and his colleague, Professor of Medicine Dr. David Aronoff, were among several Vanderbilt faculty members who were becoming familiar faces in national media. They provided expert opinions on everything from public health policy to personal hygiene. By mid-spring, VUMC experts would appear in nearly 29,000 media placements related to COVID-19 and reach a cumulative audience of nearly 77 billion. They also provided answers to common questions about the pandemic through the university's "Ask an Expert" video series.[24] Schaffner would go on to become the most-cited COVID-19 expert in the United States. In all, Vanderbilt and VUMC researchers had begun making what would be among the world's most important contributions in the fight against the virus.

First Actions

In February, Vanderbilt's coronavirus commission began assembling some of the answers to the unexpected, complex problem of how to handle campus operations amid a coronavirus. They did so based on an evolving response plan laid out in a multipage matrix. The matrix listed four threat levels—Insignificant, Emerging, Definite, and Severe—that would trigger four broad responses that ranged from "pre-planning" to "suspend selected university operations." The matrix described the critical actions necessary in each level and laid out responsibilities for every administrative corner of the university: All colleges, schools, and academic departments. Housing and dining. Greek life. Childcare centers. Human resources. Student and occupational health. Physical plant operations. Athletics. Communications. Finance. IT. Government relations. And much more. The color-coded matrix made it plain that Vanderbilt was a small city within a city, and the university was planning for how COVID-19 might affect every square inch of it.

Communication, internal and external, was proving crucial—and exceptionally challenging—in the university's efforts. A university serves many audiences, including students, parents, faculty, staff, alumni, government officials, and sports fans, to name just a few. The amount of information Vanderbilt needed to convey was growing exponentially as the virus spread, and there were few occasions when a single message could be crafted for all of Vanderbilt's stakeholders. Compounding the challenge was the fact that each of the university's ten colleges and schools had its own audience and its own communications team. With so many moving parts and such a fast-moving situation, the danger of unintentionally creating confusion was considerable.

"Every decision had a cascade effect, and a million other decisions needed to flow from it," recalls Catherine Kozak, the communication division's director of strategic projects and planning.

To make sure the university spoke with one voice, Vanderbilt centralized all communications. Everything now came through Steve Ertel's department, and the main channel for all COVID-19-related information was the university's coronavirus website, which communications staff transformed from a single catch-all page to a comprehensive standalone site in one intense, eighteen-hour push. Vanderbilt.edu/coronavirus was now the single source for information about the virus's impact on campus life; every email, intranet story, and departmental website having to do with the outbreak pointed back to the site.

"What formed our communications plan at the core were values like trust and transparency," Kozak says. "It was important to us to be clear about what we didn't know."

It was about that time, in mid- to late February, Ertel says, that the experience of responding to COVID-19 became, for his team and for personnel in departments throughout the university, an around-the-clock, seven-day-a-week "blur"—one that would stretch for months.

At the end of February, as Vanderbilt's coronavirus commission continued to lay out plans and consider the what-ifs, university leaders were most concerned about nearly five hundred Vanderbilt students studying at universities and other programs in dozens of countries around the world. Administrators compiled a list of where students were and then went country by country, attending first to students in nations where there were serious outbreaks. Eight students who had started programs in Shanghai and Beijing had been relocated to other programs in January. Nearly fifty more were studying in Italy. Vice Provost for Academic Affairs and Dean of Residential Faculty Vanessa Beasley, along with members of the Global Education Office, worked with international host organizations, as well as schools and colleges across the university, to ensure that these students were able to sustain their academic progress despite the sudden changes. Three days after the university offered guidance on the students' potential return to the US, many of the host programs in Italy started to ramp down.

Andrea Bordeau, Vanderbilt's global safety and security manager, began a string of sixteen-hour days that would stretch for more than a month without a single day off as she helped each student abroad weigh their options, and, if they chose to, return home. Every country's situation and travel regulations were different and were changing by the hour, making every student's flight home a puzzle to be solved. When Morocco suddenly closed its airspace, potentially stranding students in that country, Bordeau and Vanderbilt's travel agents monitored airlines' reservations pages around the clock so they could grab seats that became available on the last flights out of the country. Students on the Polynesian island of Samoa had to be routed home through New Zealand, more than 1,800 miles away—until New Zealand abruptly prohibited travel through the country, nullifying students' tickets. And students unable to get flights out of Paris were phoned in the middle of the night and told to sprint for a bus to another city, where flights were, at least at that moment, available.

Not all students were eager to abandon life abroad. One group of graduate students waited too long to try to return home and ended up stranded for weeks. Several other students opted to ride out the pandemic in New Zealand.

"We always had a few cases where students would say, 'You know, I think I'm just going to stay here and do my thing and I'll be fine,'" Bordeau says. "And we're left in a position of having to try to explain, 'Well, what if someone in your family gets sick? What if you have a health-care crisis of your own? Now is not a good time to get appendicitis.'" She tried to gently stress to students that, if cities locked down or borders closed, they could be beyond Vanderbilt's help.

"Vanderbilt has an international security provider that I worked with really closely—the sort of group you call when nothing else works," Bordeau explains. "And I asked, 'Can we get a helicopter to fly in and get them out?' And the answer is no. In these cases, you get to a point where there's no way I can help you. And you don't want to get there."

In all, the virus was presenting challenges unlike others Bordeau and her colleagues had faced before.

"I had worked through Ebola and through a coup in Thailand. I evacuated students out of Egypt in the middle of Arab Spring. And a global pandemic is just totally different," she says.

As Bordeau worked through the roster of students studying abroad, the university was also prohibiting travel to countries on the rapidly growing list of nations that the CDC had designated Level 3, for "high numbers of COVID-19 cases." Beyond that, approval from a vice chancellor was now required for faculty or staff to travel anywhere. Anyone returning to Vanderbilt from a Level 3 country was required to isolate—a requirement put in place more than a week before the CDC would make the same recommendation. Vanderbilt also advised international students, faculty, and staff not to travel outside the US for fear that they could get caught up in rapidly emerging travel restrictions and resulting visa complications.

"We had two lists," Turton remembers. "We had the list of countries you couldn't visit and the list of countries that, if you visited, you might not be able to return from if you weren't a resident or citizen of the US."

Wente shakes her head now at the finger-in-the-dike hopefulness behind the idea that if the university—and the nation—just managed travel, it could keep COVID-19 away. "That was the only strategy we were being offered: 'Keep it out of the US!'"

Next, administrators anxiously turned their attention to anyone planning to travel abroad during the university's week-long spring break, which would begin on February 29. Students were instructed to notify the university if they would be vacationing outside the United States. Fiona Bultonsheen, then a Vanderbilt senior, recalls: "I got an email saying that things were going to be a little different, and to mind yourself. It was a little harrowing to go into the break with the knowledge that things were happening that I didn't have control over."

Wente was conscious that other universities were watching how Vanderbilt navigated its break.

"We had the earliest spring break in the country, and that put us, in some ways, in a leadership position," she says. But as Wente and her colleagues had their eyes on students headed overseas, a threat far less expected—and more immediately devastating—descended from the skies.

Storms Converge

At the start of spring break week, in the late evening and early morning of March 2 and 3, a savage and fast-moving storm system loosed no fewer than ten tornadoes upon Tennessee. Most hit the middle part of the state, and several hit in and around Nashville—including one measuring a four out of a possible five on the National Weather Service's Enhanced Fujita Scale. All told, according to *The Tennessean*, it was "one of the strongest and deadliest storms the country has seen in years." The newspaper reported: "The vicious storms unleashed winds that spiraled up to 175 mph. They swept away homes and took away lives. They left a 100-mile scar from west to east, destroying precious pockets of Nashville and decimating the quiet community of Cookeville. Twenty-five people were killed, including five children."[25]

Shortly after midnight on March 3, the university activated its emergency operations center and notified the Vanderbilt community of a tornado bearing down on campus. The twister quickly changed direction, sparing university residence halls and facilities. But some Vanderbilt staff and faculty, whose houses were in the tornado's path, were not as lucky. Several were among the hundreds of area residents who lost their homes.[26]

As the storm passed, university managers contacted their teams to check employees' well-being. The next day, a website was set up so employees whose homes had been destroyed could report in. Eric Kopstain arranged for the university-owned Holiday Inn on West End Avenue to house any Vanderbilt employees needing shelter.

Kopstain later likened his team's initial response to the tornadoes' wreckage to that of soldiers after an attack.

"It's all just about making sure everyone's okay," he says. "It's, 'We've got something before us, and we're just going to deal with it as quickly and as effectively as possible.'" He was thankful no one in the Vanderbilt family had lost their lives. "It was bad," he says, "but not as bad as it could have been."

As news of the destruction spread, Commodores from throughout the city, across the nation, and around the world reached out to see how they could help. University employees showed up in hard-hit neighborhoods with work gloves and chain saws to help their colleagues and others. Vanderbilt Athletics hosted a donation drive before a men's basketball game. Some students changed their spring break plans at the last minute to assist in the cleanup and to help hand out meals and supplies.[27]

Kopstain recalled what it was like to be confronted with tornadoes while simultaneously scrambling to respond to the growing threat of COVID-19: "It was like, 'Are you kidding? We're about to have this global pandemic come to our doorstep, but, oh, wait, we have this big tornado to deal with.'"

For its part, the virus allowed no pause.

The United States' first coronavirus death was reported near Seattle on February 29. On March 5—the Thursday of spring break week and two days after the tornadoes—a Tennessee man tested positive after attending a conference in Boston. The event, sponsored by the biotech company Biogen, would eventually earn the distinction of "super-spreader" after tens of thousands of infections globally were traced to it.[28] On the same day, *The Vanderbilt Hustler* reported that a junior who had been studying in Florence had tested positive and was hospitalized in Chicago with mild symptoms—the first Vanderbilt student known to be infected. The student told the paper he was not the only Vanderbilt student in the Florence program who'd fallen ill.[29] University officials required everyone returning from the program to quarantine at home.

Wente notified the Vanderbilt community in an email that afternoon, noting that there were no confirmed cases on campus. "We are working with public health officials to take appropriate precautions," she wrote. "The university will resume its normal schedule as students return from spring break on Monday."[30] The university's vice chancellors, vice provosts, and deans began meeting daily.

Board of Trust chairman Bruce Evans remembers March 5 being "the real four-alarm fire bell." It was that afternoon that a member of MIT's Board of Trust, who was also the parent of a Vanderbilt student, emailed Evans. They'd seen Wente's announcement and wondered if Vanderbilt ought to be doing more. "That was the moment when it felt real," Evans says.

Wente assured Evans that Vanderbilt's response was in step with its peers—and in some cases ahead of them—in a situation that was changing minute to minute. He and Wente would be in touch almost daily for the next several weeks, and Wente would occasionally update the board's executive committee. Evans and his fellow trustees did not insert themselves into the university's decision-making.

"Things were happening so quickly, and we had such a good team led by Susan, I felt like our job at the board was to listen, to understand, and to mostly stay out of the way as the team on the ground made the emergency decisions they had to make," he says. Within two weeks, the impact of the virus's spread would hit Evans in an even more immediate way; his daughter came down with COVID-19.

By now, administrators all over the country were doing the same dreadful calculus: How much might the coronavirus have already spread on their campuses, and when was the right time to take the huge and largely unprecedented step of either locking the campuses down or—even more unimaginable—sending students home?

Universities Respond

The dominos began to fall. The University of Washington–Seattle, near where the first reported COVID-19 case in America had surfaced in January, canceled in-person classes on March 6. So did several other colleges in the Pacific Northwest. Around the country, Stanford and Rice were among the schools that followed suit.[31] Unofficially, word had come to Vanderbilt that MIT was likely to suspend in-person classes when its students were released for spring break later in the month.

At the time, the actions struck some as coming too soon. Since early in the COVID-19 crisis, Wente had been taking part in weekly calls with provosts from other private research universities to share information and best practices. When the first universities to switch to online classes announced their plans, Wente says the reaction from the provost group was, "Oh my gosh, why in the world are they doing that?" It seemed premature, based on the local environments and the information the university was getting from the CDC.

Some on the coronavirus commission argued for telling Vanderbilt's students not to return from spring break. But in the end the commission decided that, with much of the university's attention and resources focused on tornado recovery, it wouldn't be possible to coordinate such a massive, last-minute change. The decision was complicated by a lack of information about a virus whose widespread presence in the US would only become apparent in the weeks to come. With students already on their way back to campus, administrators had to make the call without the benefit of a full understanding of the risk.

"Many people later said, 'Why in the world did you let students come back to campus?'" Wente remembers. "Well, one, we were the first out of the gate among colleges releasing students for break and having them return. Two, the first case in Tennessee wasn't until late in the spring break week. Three, because many students were already returning to campus. And of course the tornado that happened demanded immediate attention also."

"Things were so fast-moving that between the time everyone went on break and came back, so much changed," Dawn Turton says. "It wasn't even day by day. It was hour by hour."

"I know the torture Vanderbilt leadership went through during those days leading up to the return from spring break, trying to manage the situation effectively with only as much information as they had," Andrea Bordeau says.

Not that the administration had taken its eyes off the virus. Across campus, administrators, staff, and faculty continued preparing for the evolving threat, working from the university's pandemic matrix. Although no one on campus had yet tested positive, plans were being developed to provide remote learning for any student who had to isolate or quarantine—and for all students, should classes go online. The Student Health Center and Occupational Health Clinic were monitoring anyone who'd recently returned from a CDC Level 3 country or who had been in close contact with an infected person; both teams were communicating regularly with Tennessee's Department of Health. Governmental rules and guidance were changing so quickly that the Office of the General Counsel, whose personnel were embedded in all the university's COVID-19 working groups, monitored changes from the federal, state, and local governments hourly to assess new policy and convey the implications to teams across campus. All the while, the Division of Communications was racing to keep up with emerging news and decisions by the university.

"We were all working insane hours," recalls Executive Director of Digital Strategies Lacy Paschal. "Up until after the end of the semester, it was not unusual for me to be on the phone with Steve Ertel at eleven o'clock at night because an email had just gotten approved and we needed to send it out to the community that night. And that was the usual. Most days of the week, we were working early mornings and late nights because things were happening so quickly. I would try to go outside for a walk for mental health, and I'd get to my sidewalk and something would have happened, and it was like, 'We need to get this message out right now.'"

The School of Nursing collaborated with campus housing managers to put up informational posters reminding students and others how to avoid infection. Off-campus housing, where students could quarantine while awaiting test results, was identified. The Vanderbilt Campus Dining team worked with their suppliers to source items that might be needed in the event of a lockdown, factoring in the scarcity of certain items if supply chains were stressed. As a backup, the team purchased thousands of military MREs (Meal, Ready-to-Eat), which have a shelf life of up to three years without cold storage. The Department of Student Athletics, like athletic departments and conferences around the country, was looking hard at how to adapt practices, conditioning, and games in the face of a wider outbreak. And the Business Services and Human Resources departments were putting measures in place to ensure business continuity if workers could not come to campus.

On March 7, Dean of Students Mark Bandas sent a message to students and their parents requiring that anyone with direct contact with someone infected with SARS-CoV-2, or anyone who had traveled to China, South Korea, Iran, or Italy, report it to the university through a special website and quarantine for fourteen days before returning to campus. Students failing to take those steps were subject to "immediate disciplinary measures."[32]

The same day, Vice Provost for Faculty Affairs Tracey George sent a message to faculty describing how the university might move forward "in the event of a significant disruption to instruction."[33] George encouraged faculty to create contingency plans that would "accommodate students or instructors who must undergo self-isolation or any other prolonged absence." The email included links to resources for "hybrid" teaching that employed in-person and online instruction.

On March 8, the Sunday before classes were scheduled to resume at Vanderbilt, Nashville Mayor John Cooper announced the city's first confirmed SARS-CoV-2 infection.[34] In response, Bandas sent another message to students, emphasizing that there were no confirmed cases of COVID-19 on campus and asking for cooperation in following safety protocols and travel guidelines.[35] The virus seemed to be closing in. It was clear that one of the worst-case responses in the pandemic matrix—"prepare to suspend university operations" or "suspend selected university operations"—would soon be necessary.

Making the Tough Call

In Kirkland Hall, where Vanderbilt's highest-level administrative decisions are made, the prospect of sending students home was the topic of intense focus. There were the practical considerations, such as quickly evacuating thousands of young people from their residence halls and the challenge of quickly switching to teaching online. And there was the fact that Vanderbilt's educational approach was rooted in the concept of residential learning.

Of equal concern were the university's researchers. Vanderbilt is one of the top research universities in the world. At its eighty-five institutes and centers, and in labs across campus, faculty and students were engaged in vital pursuits across dozens of disciplines that could not simply be abandoned without great expense, significant scholarly setbacks, and, in some cases, the loss of months or years of work.

But the pandemic seemed to be growing by the hour. And at that point, much about the virus was still unknown—and frightening.

"We didn't have a lot of knowledge about how it was spreading," Wente recalls. "We didn't know a lot about how serious it might be across different age groups and demographics. Even if you knew then that masks were important, you couldn't buy a mask anywhere, and you wouldn't want to because you'd be taking them away from health-care providers."

"People were scared," Eric Kopstain concurs. "They knew there was this virus. They knew it was highly contagious, and they knew it could kill you. How it was spread, just how contagious it was, and who might be the most susceptible—that was [to be learned] in coming weeks." At Vanderbilt, at least one student already was circulating a petition demanding that classes be canceled.[36]

Ultimately, the university's decision came down to a basic question.

"'What are we going to do to be sure our people are as safe as possible?' That was the fundamental guiding principle," Wente recalls. "I remember the Board of Trust telling me very clearly, 'Spend what you need to spend. Take care of your people.' I had a mandate to do what I needed to do."

In a March 8 email to Bruce Evans, Wente wrote, "On a daily basis, the situation changes and leads us all to question the ability to limit the impact on campus."

On March 9, Vanderbilt's campus bustled with students and faculty fresh off of spring break. And on that day, Wente and other university leaders learned about a group of nine Vanderbilt students who had spent the break visiting friends studying in Spain. One of the students, who had gone home to New York after the trip and not yet returned to campus, had tested positive for the virus. The other eight were back at Vanderbilt.

According to the pandemic matrix, select university operations were to be suspended if criteria were met indicating that the pandemic had evolved from an "emerging threat" to a "definite threat." Among them: large clusters of cases in the US, warnings against international travel, screenings of international airline passengers, fear of "worried well" Vanderbilt students and employees using local health-care resources, and a confirmed case of COVID-19 on campus. Some of the criteria had been met. The university's leadership did not wait to meet all of them.

"We pulled the rip cord," says Wente, "because you could see them coming."

"We had to get control of this because we had these students come back that we knew were potentially exposed," she recalls. "We didn't know how many might be positive. We didn't know how seriously ill they would get if they were positive or how far they might have spread it."

Wente did what she and her colleagues hoped to avoid: she announced that classes were canceled for the week and required future classes to take place online "at least through March 30."

Biochemistry undergraduate Minna Apostolova was in an evening lab for a physics course when her phone lit up with the news of Wente's announcement. Her professor told students to return to their residence halls. "That was when I think we all realized how much this was going to change things," Apostolova recalls.

Although students hadn't been asked to leave campus, Apostolova's mother back in Tulsa, Oklahoma, "saw it coming" and booked her daughter on the next flight out of Nashville.

"I packed up the two suitcases in my room with as much stuff as I could fit—the essentials. And then I flew out . . . at 7 a.m. that Tuesday," Apostolova remembers.

Late on the evening of March 9, Wente sent an email update to Bruce Evans. "I am hopeful, always the optimist, that we will have no new on-campus cases in the next two weeks and will be able to decide about re-starting in-person classes and events after March 30," she wrote.

It was not to be. Two days later, a VUMC health-care worker tested positive, and Wente had no choice. She announced that Vanderbilt would move fully to online learning for the remainder of the semester. What was more, students who lived in the US were to pack their belongings and evacuate campus within four days. For the first time in its nearly 150-year history, Vanderbilt's campus was all but shutting down.

"This is not how any of us wanted this semester to proceed, but we are acting now to ensure the health and safety of all members of our community," Wente wrote. "While our campus is the heart of this university, our people are its soul. This challenging situation is necessitating a temporary change to how we deliver education, but it will not deter us from the pursuit of our mission."[37]

On the same day that Wente made the announcement, the World Health Organization officially declared the COVID-19 outbreak a pandemic.

For many on campus, the decision to move online was stunning, even if it was not unexpected.

"In February, faculty hallway chatter was about how, once spring break comes, the students are all going to get this virus and there's going to be a disaster," remembers Associate Professor of Sociology Shaul Kelner. "But it seemed very far away. There was a sense that this was going to be a calamity, but everyone was just going about their business . . . And so when it hit, it was more shocking than it should have been, given the fact we had been talking about it for a while."

Some Vanderbilt students responded to Wente's announcement by moving their traditional St. Patrick's Day parties, thrown at off-campus apartments and on Vanderbilt's Greek Row, up by a week. One reveler appeared on social media in a hazmat suit, a mask, and a green bowler.[38] The parties turned out to be "a big spread event," Kopstain recalls.

Not all students were celebratory. Alex Livingstone, a senior studying economics and history, told *The Tennessean*: "No one really knows what's going to happen with our class schedules or graduation. For the seniors, it's especially frustrating, because these are the moments we've worked so hard for."[39]

Between March 11 and March 15, thousands of students climbed into cars and boarded airplanes bound for their hometowns. Some left campus with what they could hastily cram into a suitcase or backpack, leaving behind a living snapshot of college life circa March 2020: clothes, books, music systems, guitars, golf clubs, bedding, posters, photographs, shampoo and toothpaste, laundry supplies, and all manner of recreational diversions—university-sanctioned and otherwise. Many packed their belongings before leaving, using some of the fifteen thousand cardboard boxes, hundreds of rolls of packing tape, and bales of packing material that the university's Plant Operations team had purchased as Wente's announcement was being drafted. What would happen to students' belongings would remain an open question for several more weeks. Indicative of the level of caution regarding the spread of the virus were instructions to departing students for putting their room keys in envelopes, which politely requested, "Please do NOT lick the envelope reclosures."[40]

Like other international students at Vanderbilt, first-year political science and computer science double-major Safa Shahzad was faced with the choice between going home at a moment's notice or riding out the semester on a largely deserted campus. Given the rapidly changing situation, and with borders slamming shut around the world, she decided to fly back to her home in the United Kingdom. The day after she flew out, she remembers, all flights to the UK from Nashville were canceled. Shahzad was glad to have a seat on a plane, but she was sorry to leave.

"I was having a great time, doing way better socially and academically than [in] autumn because moving to the US as an international student when you've never been to Nashville and don't know anybody is really daunting, and I was just not in a good place," she later told *The Vanderbilt Hustler*. "And then spring was way better, but then it was cut short, which was just saddening."[41]

Some students from the US asked to remain on campus. While in the past the university had liberally accommodated such requests for Thanksgiving or winter break, this time it had to take a harder line and distinguish between students who needed to stay—for financial reasons or because of a detrimental home life—and those who simply wanted to.

"Historically, it's been, 'When we're open, we're open,'" says Senior Director of Residential Experience Randy Tarkington. "I always joke that we're like Motel 6: the lights never go off. But for the first time, a student saying they wanted to stay wasn't enough. It wasn't about convenience; it was about safety, and we had to de-densify the campus. That led to days and days of really tough conversations."

As students made their way home, Vanderbilt's faculty pivoted quickly—with the help of the IT team—to provide instruction online using a combination of lectures streamed during normal class times and videos and other content that students could use on their own time. Vanderbilt's Center for Teaching offered workshops to help faculty quickly translate their in-person classes online. The university's IT department expanded its hours of support and developed web-based resources for faculty. They stocked up on laptops to have plenty on hand in advance of a projected global computer and webcam shortage. In addition, they increased the capacity for phone support by installing 368 new lines and establishing dedicated fiber connections to support the increased number of calls to VUMC. By March 12, the IT team had obtained about twenty-four thousand licenses for Zoom, the soon-to-be-ubiquitous application that locked-down Americans would use to work, go to school, and stay in touch with loved ones.

Vice Chancellor of Information Technology John Lutz experienced the pivot from two perspectives—as the head of IT and interim head of development and alumni relations, and as an instructor in Vanderbilt's Owen Graduate School of Management, where he teaches a course in IT for business leaders.

"On March 7, I returned from spring break with my kids. On March 9, I taught my first and what would turn out to be my only in-person class that semester," Lutz recalls. "I delivered the rest of that module via Zoom from a hastily organized home studio, all while taking on the new challenges in IT and donor and alumni relations. It was a wild but fun and inspiring ride to see all of Vanderbilt rise to the occasion."

In all, faculty moved with remarkable speed to create a robust online environment in which students could continue their classes and successfully conclude the semester with full credit for their courses.

Yi Ren, an assistant professor in the Department of Biochemistry, remembers a general uncertainty about what would happen next. As a biologist who knew something about viruses, she says she was "less likely to panic than most." Still, when the news came that classes would move online, she was glad.

"It was beyond anyone's imagination," she remembers. "There were still so many questions about how COVID-19 could be transmitted."

Mechanical engineering assistant research professor Kevin Galloway suddenly realized he and his colleagues were going to have to look at their work-life balance in a whole new way. Galloway had just taken his child out of day care. Now he'd be working from his house. "We were trying to balance everything under this heightened sense of uncertainty," he remembers.

Faculty would not be alone in merging home and work. On March 13, "we sent all the nonessential staff home [to work by remote]," Eric Kopstain recalls. It was supposed to be "just for two weeks, because we didn't know what was going to happen. That's so laughable now. But we were trying to figure out what to tell people. We didn't want to create mass panic."

On March 15, the university recommended that students still studying abroad—in countries including Spain, Denmark, and Australia—return home to their permanent residences. Research operations also began to wind down. The university restricted all on-campus research laboratories to "essential activities only."

Two days later, Vanderbilt's Board of Trust announced its Ad Hoc Committee on University COVID-19 Response. After the board had largely empowered Wente and the rest of Vanderbilt's leadership to do what needed to be done during "an incredible seven days of crisis," Bruce Evans was now forming the ad hoc committee "to keep the board informed and involved in what lies ahead." The panel would be headed by Vanderbilt alumna Nora Wingfield Tyson, a retired vice admiral and former commander of the US Navy's Third Fleet. The nine-person committee also included Evans, Wente, Balser, incoming chancellor Daniel Diermeier, and trustees Douglas Parker, chairman and CEO of American Airlines Group; Jeffrey Rothschild, former vice president of infrastructure engineering at Facebook; Alex Taylor, president and CEO of Cox Enterprises; and Ike Lawrence Epstein, senior executive vice president and COO of Ultimate Fighting Championship.[42]

It was also on March 17 that the Southeastern Conference, one of college sports' "Power Five" leagues and the conference to which Vanderbilt belongs, canceled all spring sports competition—including the annual spring football scrimmage and the

SEC Baseball Tournament. Football "pro days" for the conference's athletes who had declared for the 2020 NFL Draft were also prohibited.[43] Vanderbilt's athletic department followed the SEC's lead. All other activities for in-season sports were canceled until at least mid-April; all out-of-season sports were suspended until further notice. Wente had called SEC commissioner Greg Sankey directly to tell him the university had sent students home.

The loss of the spring season was a blow to Vanderbilt women's tennis player Christina Rosca, the top-seeded member of her team. A senior, Rosca was convinced her college tennis career had come to an end.

"I thought I was done," Rosca says. "I was going out of my college career, which was sad because I had invested so much time and effort, and [we] were having so much fun. So that was all very, very difficult."

The decisions to send students home and cancel the spring season had come just weeks after Candice S. Lee had been named Vanderbilt's interim athletic director. (The Vanderbilt alumna, former women's basketball captain, and longtime athletics department staffer would be named to the position permanently in May, making her Vanderbilt's first female AD and the first African American woman to head an SEC athletics program.) There was no precedent to guide Lee through protecting her student-athletes during a global pandemic. She found herself trying to answer questions that had no immediate resolution: What would the shutdown mean for student-athletes and their eligibility? Would student-athletes return in the summer as usual? What did all this mean for incoming first-year students?

"The most challenging part was communicating as quickly as possible, knowing you do not have all the information," Lee says. "You're trying to keep people updated and involved in real time as much as you can. That was really hard. There were so many questions, and we didn't know what the end date was."

While most student-athletes had packed up and headed home, others had special circumstances that Lee needed to accommodate. Some were international students. Others faced challenging home situations, involving everything from poor family relationships to a lack of internet service to not having enough food.

"These are the intricacies that are surprising to some people, but all of a sudden you have to figure out how to adapt," Lee says. "A lot of times I would take the words of Greg Sankey and do my best to apply them because I thought he did a great job in leading us. He would say, 'Listen, you all understand the challenge, right? We're building this bridge at the same time we're crossing it.'"

The Campus Goes Quiet

In all, the first two weeks of March were a flurry of decision-making, with leaders across the university making their best calls based on a constant flow of new information and little to no specific guidance from state or federal officials. Of course, Vanderbilt wasn't alone. More than 1,300 colleges and universities across all fifty states ultimately canceled in-person classes or shifted to online-only instruction.[44]

Yet for all the action and speed, the wind-down was more or less proceeding according to plan.

"We were, in an orchestrated way, using the pandemic matrices we had prepared leading up to this," Kopstain explains. "Very quickly, we went from full operations to very highly curtailed operations. Our planning served us well."

Zoom played an important role, enabling dozens of homebound managers from across the university to meet.

"It was very tactical, operational stuff," Kopstain recalls. "We'd say, 'Okay, where are we on packing up student belongings? Has everybody left yet? Have we locked all the buildings down? Are the day cares closed?'"

By the middle of March, Vanderbilt's campus was still. Yet it never fully closed. Several hundred students, including many international students and students who simply could not return home, remained in their residence halls. Faculty who lived permanently in the residential colleges remained in their homes. In all, about five hundred people would stay on campus through the end of the semester. That meant some dining facilities and residence hall workers stayed on the job too. Meal service was consolidated in two dining halls where students could pick up boxed meals from tables outside. A system was set up so that students isolating or quarantining could order meal delivery by text.

Other essential staff, who were responsible for building maintenance, public safety, and more, showed up every day—adapting their jobs to the realities of COVID with social distancing, constant sanitizing, and, when public health officials began recommending it, mask-wearing. At each step, the university used all available information to adjust its protocols to keep workers and the broader campus community as safe as possible.

At the same time, researchers at VUMC continued to step up in the fight against the pandemic. Moderna began human testing of its promising mRNA-1273 vaccine in mid-March; Mark Denison led Phase 1 testing and would go on to lead Phase 2 clinical trial analysis and reporting in May.[45]

Denison and his team had also solved a big problem with more local implications. Initial coronavirus testing kits distributed by the CDC had proved unreliable, so Denison's lab worked with the VUMC pathology lab to develop their own test. By the middle of March, VUMC was able to test seven hundred to eight hundred people daily, accounting for three-quarters of the testing happening in the state and providing the data set that informed Metro Nashville's March 15 decision to close bars, restrict restaurants to take-out service, and take further precautions in an effort to help slow the spread of the virus.[46]

On March 20, the university announced it would refund housing and dining fees to students who had left campus. It also launched a Student Hardship Fund to provide financial support for those experiencing pandemic-related difficulties.

Three days later—nearly two weeks after Vanderbilt ramped down on-campus activities— Nashville Mayor John Cooper and the Metro Public Health Department ordered residents to stay in their homes except for essential activities.[47] The same day, Wente established a Public Health Advisory Task Force to develop the university's COVID-19 protocols and best practices.

"We are fortunate to have public health and infectious disease experts on our faculty," Wente said in announcing the task force. Notably, she added, the task force would advise on "reversal triggers" for reopening research labs, bringing staff back to campus offices, and restarting in-person classes. Underscoring the School of Nursing's essential role in the university's response to the pandemic, Dean Linda Norman was named chair of the task force.[48]

Wente also announced the creation of a University Continuity Working Group that would collaborate with the task force to advise on operating safely in the months to come. The panel was divided into smaller working groups, each charged with making recommendations in one of four areas: education continuity, research and scholarship continuity, business continuity, and community continuity—the latter of which would explore ways to keep staff engaged while working remotely. Each group would be composed of faculty, staff, and, in some cases, students. The creation of the working group and the task force signaled that administrators were already considering how to fully reopen, even as the lights on campus dimmed.

In February, the global momentum of the COVID-19 pandemic had felt like "this big wave that was starting to form," remembers David ter Kuile, executive director of campus dining and business services. By the end of March, the wave had crested and broken. Sending students, faculty, and staff home had been a drastic but necessary step. Like a growing share of the nation, Vanderbilt now was hunkered down and waiting to see how the rest of the pandemic would unfold. As spring came to Middle Tennessee, the campus was perhaps as quiet and empty as it had ever been. But the coming fall? If Vanderbilt's leadership had its way, that would be a different story.

PART 2

THE ROAD TO REOPENING

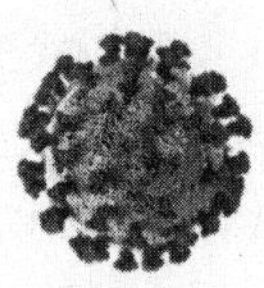

"Vanderbilt University is a center for scholarly research, informed and creative teaching, and service to the community and society at large. Vanderbilt will uphold the highest standards and be a leader in the quest for new knowledge through scholarship, dissemination of knowledge through teaching and outreach, and creative experimentation of ideas and concepts. In pursuit of these goals, Vanderbilt values most highly intellectual freedom that supports open inquiry, equality, compassion, and excellence in all endeavors."

—VANDERBILT UNIVERSITY MISSION STATEMENT

The first woman to give the chancellor's commencement address at Vanderbilt University did not deliver it from a stage on Alumni Lawn on a glorious May day. She was not looking into the exuberant faces of thousands of seated graduates in black caps and gowns. There were no families watching with pride and relief, no faculty in hoods and tams, no brass quintet from the Blair School of Music.

Instead, Susan R. Wente gave her address to a video camera, carefully lit and reading from a teleprompter in the Great Room of E. Bronson Ingram College, with only a socially distanced camera crew and a few aides to hear her. There was no pomp but plenty of circumstance.

Not only was Wente, as interim chancellor, the first woman to give the address at Vanderbilt since commencement ceremonies began in the nineteenth century, it also was the first time in the university's nearly 150-year history that the commencement celebration had been canceled.

The decision to cancel had been made in March, not long after students had been sent home. Vanderbilt's leadership had first considered postponing the ceremony—perhaps by a few weeks, or maybe until the fall—as some other universities were doing. But with large gatherings banned on campus and the trajectory of the pandemic unclear, organizers—with input from students—ultimately elected to plan an event that would serve as both ceremony and reunion for the Class of 2020 in May of the following year.

Nothing on the university's calendar carries the emotional weight of commencement—an event that is at once a celebration, a farewell, an acknowledgment of years of work, and a life milestone for students and their families alike. Its cancellation brought home the reality of COVID-19.

"When it became clear that commencement wasn't going to happen, that was the wake-up call that our world had changed," says Vice Chancellor for Finance Brett Sweet.

Wente had broken the news to students on March 25, in writing and by video.

"This breaks my heart for each and every one of you," she told them. "But it's the right decision, and the one that's guided by our strong belief that we must do our utmost to protect our beloved Vanderbilt community, our newest graduates, and their supportive and loving families."[49]

Looking back months later, Wente says that while forgoing commencement was wrenching, the actual decision to cancel the event was easy to make. It was, in fact, the only choice given the worsening pandemic.

"It was obvious," she says. "No way were we going to be able to have commencement. Nobody was kidding themselves. But among all of our peer universities, who wanted to be first to cancel?"

Efforts began in April to create a virtual commencement, including a special Class of 2020 website, congratulatory videos from faculty and staff, special Zoom backgrounds for graduating students, and more. When the original commencement date of May 8 arrived, Wente, in the video she'd recorded, officially conferred graduates' degrees, which would be delivered by mail. With palpable emotion, she gave graduates her benediction:

> Through innovation and a measure of grace and good humor, you have shown the world what resilience looks like. You have learned the painfully persistent truth that life is not always predictable, easy, or fair. And yet, through all of these trying circumstances, you have had to remain steadfast in your work.
>
> . . . None of us could have predicted what you would experience as your final spring semester. This is a generation-defining moment. You are the class that future and past Vanderbilt alumni will look to in awe, recognizing the sacrifice you were forced to make at the culmination of your educational journey at Vanderbilt.
>
> . . . I hope you see how your Vanderbilt education—the skills and habits that you've developed, the understanding of the world that you've gained—can serve as a source of comfort and resourcefulness and optimism during these hard times. I hope you have learned that nothing is so daunting that it cannot be explored, learned from, and ultimately overcome.
>
> . . . This being Nashville, if I were to play a song for you right now, it would be the words written by Mark Sanders and Tia Sillers and sung by Lee Ann Womack: "When you get a chance to sit it out or dance, I hope you dance."

Following commencement, each graduate was mailed, along with their degrees, a VU champagne glass and a note from Wente. They would convene with her online in June for a virtual toast.

Going without the closure of a traditional commencement was difficult, says Fiona Bultonsheen, a 2020 graduate in economics. "I was able to go back to Nashville in the summer to grab the rest of my stuff and say hi to some friends. But it did feel like we all, very suddenly, had to move on to adulthood."

Bultonsheen captured her feelings about the semester in a bittersweet song she and a classmate wrote in an online collaboration with their professor in a songwriting seminar at the Blair School. It included these lyrics:

They stripped everything and now it's really tough
And every day is just another day closer to enough
I'm on a stay-at-home vacation
But man what happened to my life's celebration?[50]

Students weren't alone in mourning commencement.

"Graduation is one of my favorite times of year," Vanessa Beasley says. "I sit in the front row and take as many photos as I can on my phone and then mail them to students the next week so they and their families can have the unexpected candid shot of the moment before they crossed the stage." An encounter Beasley had with a departing student in March, who shared with her how he had tried coming to terms with the unceremonious end of his time at Vanderbilt, she says, "forced me to start processing the loss of that ritual in May and what it meant to me too."

"I definitely wished there was a better ending," Bultonsheen says of the semester. "But the world has really changed in so many other ways that, for me, the lack of commencement isn't the worst thing that has happened in the past couple of months. . . . Having any sort of gathering at that time would have really been a mistake."

Finishing the Spring

In the weeks between students leaving in March and the day Wente taped her speech, Vanderbilt's remaining on-campus staff had been busy cleaning up after the abrupt student move-out and running a partly open campus that was housing a fraction of its usual spring semester population.

One task was determining what to do with the tons of belongings students had been forced to leave behind when they moved—some of it packed in cardboard boxes, some not. Shipping the belongings to students would be expensive. Storing them presented its own challenges, starting with figuring out where.

The university's decision was forced when VUMC, planning for a possible surge in local COVID-19 cases, requested that a thousand rooms be cleared for patients. Rooms were also set aside for health-care personnel who were leery of commuting between work and home and potentially infecting their families. To clear and prepare the rooms, facilities teams, who were dubbed "The Ghostbusters" for the tanks of disinfectant on their backs, first went building to building, room to room, on campus and off, spraying down all surfaces as a precaution for the moving crews who would come behind them.

"We didn't know enough about the virus at that point to understand that if we had just left buildings sitting for seventy-two hours or so, the virus would have died off on its own," Assistant Vice Chancellor for Plant Operations Mark Petty says now. The university was working to keep its people as safe as possible, based on the information available.

Before students' items could be stored, they had to be inventoried—and there were thousands of rooms to go through.

Once the spaces were deemed safe to enter, crews labeled and collected the contents of each room. Occasionally, they collected contents outside the rooms as well.

"I'll never forget this: There was a team of three of us who cut 364 bikes loose from the racks, just so nobody would steal them," recalls Sports Turf and Facilities Manager Ryan Storey. "Three hundred sixty-four bikes—it comes out to almost one for each day of the year." Storey and his coworkers stored the bikes until students could claim them.

Once all the belongings were packed and labeled, an improvised caravan of pickup trucks ferried the goods to what, jokingly and not, became known as "the undisclosed location"—the Stadium Club, a ballroom-sized space atop a parking garage that looked into Vanderbilt Stadium. The big room had once been used for VIP events, then served for a time as the staff gym. Crews had lined the space with tens of thousands of dollars' worth of shelves. "It looked like the warehouse at the end of *Raiders of the Lost Ark*," Petty says.

Teams had to make up the moving and storage system on the spot because the situation had developed much too quickly for any logistics planning to be done. But it worked.

"We literally became a storage operation overnight," Senior Director of Residential Experience Randy Tarkington says.

Vanderbilt's dining services staff were improvising, too, as they determined how best to feed the several hundred people still living on campus, as well as public safety staff and other essential workers. Some painful choices were required. To minimize the possibility of exposure to the virus, about three hundred hourly employees had to be furloughed. Remaining staff members consolidated dining operations at Rand Dining Center. Separate work teams were formed so that in the event of an outbreak among one team, all members could quarantine and another team could be brought in without a disruption in service.

Other teams on campus were adapting as well. The university's IT department stayed in high demand to support remote work and online classes. The public safety team expanded its scope by providing on-call medical transportation for VUMC and increased the number of officers on duty to help monitor people entering and exiting the medical center. Vanderbilt University Child and Family Centers stayed open for VUMC faculty and staff working extra hours.[51] And faculty and grounds crews solved novel problems created by the shutdown.

Storey recalls: "When, out of the blue, they said all sports seasons were canceled, I was like, 'What do we do with all the fields?' In the end we found ways to save time, labor, and expenses—like by spraying a growth regulator on the grass to slow it down."

Working closely with the team in Vanderbilt's Office of the General Counsel, the university's Human Resources team helped workers navigate the new landscape.

"We worked hand in hand with each department to figure out what functions were still required, how we perform those functions, which people would be needed, and how to keep them safe," says Cleo Rucker, director of HR Consulting, Employee and Labor Relations.

Teams throughout the university continued to work long hours. "You'd do a fourteen- or fifteen-hour day and think, 'I just got here,'" Petty remembers. "We tried, as much as possible, to respect weekends so that people could get a break. But in the beginning, I think we'd spend half a day, or six to eight hours, on the phone or on Zoom on Saturdays, and usually a couple hours on Sundays, trying to figure out what to do next."

All the while, the darkened buildings and empty green spaces on Vanderbilt's campus gave no indication of the instruction that was still happening on computer screens as faculty and their students finished the semester while scattered across the globe. Using video, conference calls, email, and more, Vanderbilt's faculty took their courses online—more than 2,600 of them in all, with as many as 187 classes estimated to be occurring simultaneously. Over 550 faculty members had participated in the Center for Teaching's "Tools for Putting Your Teaching Online" sessions. During the first half of March alone, they uploaded more than 4,200 hours of video content to Kaltura, a multimedia tool for online courses.[52] For assistance with the high number of requests for technical help, IT staff called on faculty and the school's audiovisual teams to operate "pop-up studios" for faculty. They also partnered with staff from the Owen Graduate School of Management to build online training classrooms, where professors learned to teach remotely with various platforms.[53] In all, IT staff helped about 1,300 faculty members make the transition to teaching online.[54] Once classes were up and running, a typical day had roughly thirty thousand Commodores participating in some four thousand Zoom meetings.

Not all faculty and students were new to online instruction; those in the School of Nursing were pioneers in this area, having been a part of online instruction broadly for years. "Our faculty and staff are very tech-savvy, so we were able to avoid a lot of the technical difficulties that others might have faced," Program Coordinator Allie Noote says.

As part of a roundup of faculty insights about online teaching, Lori Catanzaro in the Department of Spanish and Portuguese reflected on the faculty's new view into students' lives. "We can see slivers of their homes through the Zoom squares," she wrote. She observed students serving as caretakers for older relatives and couch surfing in Nashville because they couldn't go home. Some families had not previously had internet access and had scrambled to get it when public libraries had closed, Catanzaro noted.

As students and faculty slowly grew accustomed to the online landscape, staff in university departments that were now spread among homes across greater Nashville got their bearings as well.

"For that first month after students moved out, our office was focused on figuring out what was happening with the students and getting them situated before we could focus on what our staff needed," says Anna Thomas, assistant director of training and communications for the Office of the Dean of Students. When her department finally had the time to consider its next steps, "it was about connection at that point. We weren't thinking about professional development, per se, we were thinking more, 'How can we keep our community intact?'" Thomas and her colleagues scheduled virtual coffee hours. "There was no agenda. They were opportunities to get to know one another under different circumstances and talk about how we were all coping."

Committed to the Fall

As Vanderbilt settled into something of a post-shutdown routine, its administrators had a brief moment to catch their breath and consider their next steps. There was really just one question on everyone's mind: Would students be back in the fall? Eric Kopstain remembers that, as early as April or May, the answer for most of the university's leadership was yes.

"As we met with the board's ad hoc committee, from the beginning, there was never any question at all, period, that our goal for the fall was going to be to have people here to the largest extent possible and reopen," Kopstain says. "There was no hesitation from our board on that. And I think we were among a small group of institutions that were resolute very early on."

Wente agrees: "The pivot to reopen wasn't very long after we closed."

By spring, incoming chancellor Daniel Diermeier was more involved in decision-making, taking part in online meetings from his home in Chicago while bringing his tenure as the University of Chicago's provost to a close. Much of Diermeier's wide-ranging résumé, which included professorships in the business schools at Stanford, Northwestern, and the Harris School of Public Policy at the University of Chicago, now read as preparation for managing a university during the COVID-19 pandemic—particularly his scholarship in crisis management. Like Wente and Kopstain, Diermeier recalls that Vanderbilt's leadership did not waver on whether to bring students back. Some of that resolve stemmed from how administrators framed the question.

"We did not approach it like, 'Should we go online or should we go in person?'" Diermeier explains. "Once you frame it that way, the easier choice is to move everybody online because inviting students back safely would require a Herculean effort. We thought about it differently. We said, 'What is our purpose? Who are we?' And our purpose is, we are a research university and committed to residential education. What that means is we want the labs up and running as quickly as possible, and we want to invite students back to the maximum extent possible in the context of the pandemic. We never said, 'Certain things are not important, so we won't do them.' We always said that whatever our faculty and students are passionate about, we're going to try to enable. So, if our students at the Blair School of Music had particular needs, we would try to address them if at all possible.

"You also have to ask, what are the consequences of not providing education? Because it's not like what we do is some luxury. We're providing transformative education, and we're doing pathbreaking research that has dramatically positive consequences for people's lives."

"The mission of the institution is not to eliminate risk. It is to provide education and support research," says Vanderbilt general counsel Ruby Shellaway, describing her thinking at the time. "What we have to figure out as an institution is what that balance looks like. And so it was a constant conversation about what the right balance was—always, always with the safety of the campus community at the front of our minds. . . . We said, 'What would the risks look like? How would we manage them? What would a successful program look like?'"

University leaders knew reopening would only be feasible if there was a comprehensive plan to keep the community as safe as possible. Vanderbilt's access to the infectious disease experts at VUMC and its long partnership with the School of Nursing put it in an especially strong position, administrators reasoned. And if infection rates rose too high before classes started, or it was proven that students could not live and learn together safely, the university could quickly revert to all-online teaching, applying what had been learned in the spring. Even as leaders began looking at how to have students back on campus, they also set out to develop an equally excellent online experience for students who had health conditions or other limitations that would require them to study remotely. The objective was to offer two robust options for the fall. This would allow instruction to change back and forth between online and in-person, as needed, depending on the current public health situation.

Of course, having a goal was not the same as having a plan. After winding down campus operations in a manner that was, all things considered, relatively smooth, "we realized that it's going to be ramping back up that's difficult," Kopstain says. So Vanderbilt's leadership, almost all of them working remotely, began reading "a ton of stuff"—articles and reports to better understand the virus and how society, including universities, could operate safely. Among the literature was a paper from the American Enterprise Institute that Board of Trust chair Bruce Evans had recommended, titled "National coronavirus response: A road map to reopening."[55]

"A little bit more was being learned about the virus, and it became clear that it was more dangerous if you were older and had health conditions," Kopstain explains. "And what happened was, people started to publish plans and said, 'Look, here's a way to reopen.' The realization was that this is not a light switch, it's a dial. It became clear around the world that it's possible to start to reopen society, but it had to be done in a very measured, careful way."

Kopstain and his colleagues looked at examples from cities around the world. They looked at Wuhan, which had finally come out of lockdown. They looked at South Korea's successful early strategies for containing the virus. They looked at New York, which had suffered an explosion of cases and deaths during March and April and was fighting desperately to flatten the curve.[56] Kopstain spent hours on the phone with anyone he could think of who might have insight, including friends at high-profile business consulting firms. Administrators across Vanderbilt were comparing notes with their peers at other universities. When the city of Nashville published its plan for reopening, Kopstain and his colleagues studied it, too, to make sure Vanderbilt's plan would comply—and to borrow any best practices.[57] The vast medical center adjacent to campus was also an invaluable resource. Activities between Vanderbilt University and VU Medical Center were closely coordinated with cooperation between the leadership of both institutions. Balser also participated regularly in the meetings of the Board of Trust's COVID-19 ad hoc committee.

"It was essential to us that we talk to the medical center and the people at the medical school," Diermeier says. "We based our decision-making on the best science and public health information. So we were constantly in communication with them. They were great partners for us."

From the start, the objective of the reopening plan was clear: it would be aimed at supporting those activities most critical to Vanderbilt's mission, including teaching and instruction (on campus and online), residential living, on-campus research and scholarship, and the administrative and operational support needed to make those things happen.

Also articulated early on were the mandatory campus-wide safety protocols that would be put in place. They included testing and tracing, symptom monitoring, mask-wearing, physical distancing, management of foot traffic in buildings and on campus, restrictions on university-sponsored travel, and more. At a May meeting of the Board of Trust's COVID-19 ad hoc committee, the university's leadership team laid out guiding principles for the reopening effort in a presentation slide:

> Vanderbilt will ramp up in phases with primary mission activities at the heart of all decision-making. The health and well-being of our community is critical, and only those who need to be and can be on campus safely will be. A data-driven, proactive approach, based on state, local, and VUMC public health guidance will drive decisions. Risk tolerance must be set at appropriate levels—neither too low nor too high. Decisions must be implemented with a philosophy of planning for the best-case scenario and preparing for the worst. Radically different ways of deploying VU campus resources, including physical space, will be necessary as part of the new normal. Many rules and protocols for on-campus activities will be established on a campus-wide basis, designed to minimize risk of disease spread while contingency plans will address potential disease outbreaks. Protocols must empower individuals to act responsibly and be an active part of the solutions, showcasing Vanderbilt as a unified community. Protocols will be pragmatic and evolve over time.

Overall, Vanderbilt's plan would be based on three pillars: follow the best available science to keep the community as safe as possible; build in flexibility, given the uncertainty inherent in the evolving pandemic; and adapt as new information becomes available.

The plan that came out of administrators' work in April and May became officially known as the Return to Campus Plan. Coordinated with the plan to reopen Nashville,[58] it was, like the pandemic matrix, conceived as having four phases—in this case, Begin, Continue, Intensify & Launch, and Sustain—that the university could toggle among as circumstances dictated. The plan governed all aspects of executing the university's mission and also laid out guidelines for preparing the campus and recalling staff. Directives under each phase were less prescriptive than those in the pandemic matrix and sometimes overlapped several phases, affording administrators more flexibility. They ranged from increasing research capacity back to 33 percent and preparing for in-person undergraduate instruction under Phase I, to "Ready for on-campus activity and/or enhanced/virtual/alternative platforms" under Phases III and IV.

The Financial Impact

One variable adding to the complexity of the university's Return to Campus Plan was that springtime at Vanderbilt is budget season. Looming over the plan to bring students back was the potential financial impact of the pandemic and the recession it had helped set in motion. It was no secret that keeping students home threatened to wreak fiscal havoc at colleges and universities across the US. As administrators everywhere looked at ways to resume in-person classes, critics pointed to the prospect of lost revenue as their sole motivator. Vanderbilt was in a strong financial position compared with many other schools, but it wasn't immune to possible revenue shortfalls.

"Once we had everyone as safely protected as possible and had gone remote-only, we had to then address ensuring that we had a strong financial platform and a plan to take care of our people," Susan R. Wente says. "How are you going to ensure that you don't have to lay a lot of people off? How are you going to ensure that you are stewarding resources in the right ways?"

The possibilities were worrying. All of the university's income sources were suddenly under pressure. Tuition made up the largest share of the university's operating revenue, but in tough economic times, more students were likely to need financial aid, which reduced net tuition revenue. (According to Vice Chancellor for Finance Brett Sweet, only about 20 percent of Vanderbilt families pay "full sticker price.") Donor support, another important source of income, typically dipped during economic downturns, as did grants supporting research. In the near term, the university had already lost the housing and dining fees that they had refunded after most students left campus, and it was possible it would forgo more if a significant number of students chose not to return in the fall. Also in doubt was athletics revenue from broadcasting, sponsorships, and other sources.

In an April 17 message to the Vanderbilt community, Wente warned that the university expected negative budget impacts of more than $30 million in fiscal 2020. "Depending on the length of the pandemic, we could incur significant additional losses in revenue and increases in costs over the next several months," she wrote. She said the university would be taking a number of steps, including suspending discretionary spending through the end of the fiscal year, reviewing construction and renovation projects, restricting staff hiring, and deferring merit salary increases planned for fiscal 2021. There would be a short-term voluntary departure program that offered compensation to a small number of staff who agreed to leave their jobs, but "based upon current known conditions, there is no plan at this time for across-the-board university layoffs or furloughs." The goal was to reduce university expenses by between 5 percent and 10 percent in fiscal 2021,[59] with expenses related to support functions, rather than instruction, being trimmed most.

The good news was that planning for an economic downturn had been done the previous autumn, when Sweet and his team had worked with Vanderbilt's deans to develop a budget playbook for a recession that many economists were anticipating before anyone had even heard of COVID-19. Vanderbilt could use that playbook as a starting point, but it would require fast action and some tough choices. "Basically, we had to redo all the budgets," Wente says.

The budget that Sweet's team had mostly completed in March and would be sending to the Board of Trust in April was now deemed provisional. Sweet—who had served as an officer on a nuclear submarine and liked to tell his team that even on their worst days at Vanderbilt they still got to see the sun shine and didn't have to worry about having enough oxygen—asked his department to prepare a revised budget for the board's adoption in June. At the same time, Wente took the unusual step of asking Anders Hall, Vanderbilt's vice chancellor for investments, to begin sending her weekly updates. Hall and his team of twenty-two people managed the 2,800 funds that made up Vanderbilt's endowment. The endowment was crucial to funding a number of programs for students and faculty, including the university's groundbreaking Opportunity Vanderbilt financial aid program, which allows qualifying students to graduate debt-free. The endowment had totaled about $6.3 billion in fiscal 2019.[60] But with stock markets diving in reaction to COVID-19, the endowment had shrunk by about $1 billion between February 20 and April 7. A long-term loss in value could affect annual endowment distributions for years to come.

The university's finance team would eventually identify about $88 million in COVID-19–related financial losses projected during fiscal 2021, owing to declines in tuition, donor contributions, investment income, and trademark, license, and royalty revenue, as well as the losses of athletics ticketing and broadcasts, among other revenue sources. Sweet warned that another $79 million could be lost if students could not return to campus in the fall. In the expense column, the university would likely be investing tens of millions to equip and adapt facilities for operation under safety protocols during the pandemic.

About $68 million of the expected losses could be mitigated through belt-tightening, hiring and salary freezes, and voluntary separation by a relatively small number of staff. The $1.46 billion revised budget Wente and Sweet would ultimately present to the board in June would be based on projected earnings of $20 million less than the provisional budget passed in April, a decrease of about 9 percent. The university would not escape the fiscal effects of COVID-19, but, for the moment, it would be able to avoid more stringent measures—such as across-the-board layoffs and eliminating contributions to employee retirement funds—that many of its peer institutions were being forced to take. Years of foresight and careful stewardship had given Vanderbilt a solid balance sheet; it was in sound condition to weather the storm.

Vanderbilt was also in a position to offer assistance to students who were struggling financially because of the pandemic. About a week after sending everyone home, the university had committed $1 million to create a student hardship fund. Students with demonstrated need could receive funds to cover last-minute airline tickets, housing, meals, online learning technology, and other necessities. More than

500 alumni donors had added over $170,000 to the fund by mid-April, and more than 1,200 students from across all of Vanderbilt's ten schools and colleges had applied for help. Among the recipients were a doctoral student facing financial insecurity because of a job loss, an immunosuppressed student needing to travel for routine medical treatments, and a student returning from working abroad.

The university also announced it would use the $2.8 million it was scheduled to receive as part of the federal Coronavirus Aid, Relief and Economic Security (CARES) Act—Congress's pandemic relief and stimulus package—to support students facing financial difficulty. To prioritize students with the greatest need, funds would go to financial aid recipients whose families were expected to contribute $10,000 or less to their tuition—or, in the case of graduate and professional students, students who received a federal student loan. Each of these students would receive $1,100. Some 20 percent of Vanderbilt's students, split almost evenly among undergraduates, graduates, and professional students, would eventually receive funds.[61]

The Return to Campus Plan

As the Return to Campus Plan came into focus, so did the vast scope of work that would be required to execute it. Much of the physical environment of the campus, some ten million square feet of space across several hundred buildings, would have to be altered. Each building would have to be mapped and signs posted to identify safe egress and ingress. Centralized testing and tracing programs would have to be built from scratch. Everyone in the campus community would need education—and reminders—about social distancing and masking. Throughout the spring and into the summer, it would take "a gigantic effort to get all that done as quickly as we could," Kopstain says.

To handle the load, multiple teams of university leaders were working simultaneously. The thirty-seven-member coronavirus commission that Wente established in January met every Friday. The board's ad hoc committee met regularly as well. A "Scenario Response Organization" was composed of ad hoc planning committees that were divided into two groups: a "mission team," which focused on restarting research, facilitating teaching in-person and online, reopening residence halls, and advising on student-athlete issues; and a "people team" responsible for assessing the impact of SARS-CoV-2 on the workforce, enabling remote work, and leading any necessary workforce reduction. The university's leadership team, composed of vice chancellors, deans, vice provosts, and other key staffers, met daily.

Central to the working groups' planning was a disciplined decision-making process. The open questions and shifting course of the pandemic—what Diermeier called "the profound and persistent uncertainty"—meant decisions would carry residual risk, and working groups needed to be fully aware that, despite best efforts, things might not turn out as planned.

To ensure sound outcomes, teams encouraged frank dialogue and employed techniques like "red team" exercises, borrowed from the field of cybersecurity, in which teams were asked to find the weaknesses or flawed thinking in existing processes, and "pre-mortems," where negative or unintended consequences could be identified in advance.

"These exercises significantly improved the quality of decision-making," Diermeier says. One example: Adjustments to the university's academic calendar seemed sound on paper but, when discussed during a red team exercise, were proven to be based on assumptions that were no longer valid.

"You could not sit with these problems . . . and then come up with good solutions based solely on your own thinking," says Vice Provost for Academic Affairs and Dean of Residential Faculty Vanessa Beasley. "You needed to be able to formulate and state your arguments and then listen to counterarguments. You needed other people to help you imagine unintended consequences. You needed to listen and be open to letting your position and your perspective change. Even then, the solutions may not be ideal, but only then could you say, and truly know, that this is the best we can do."

The intense consideration given to reopening scenarios continued despite the fact that no official decision to reopen had yet been made. Vanderbilt's leadership was still waiting to see how the pandemic would play out in Nashville and the country at large. But the university raced forward in its effort to be ready by August, in case conditions for reopening then were right.

On May 7, Nashville Mayor John Cooper announced that the city, which had been in lockdown since late March, would begin reopening four days later, despite case counts climbing steadily toward what at that point would be their peak. The same day, Wente shared the news that Vanderbilt was developing its own framework for gradually reopening, noting that Nashville's plan did not provide guidance for universities, given their unique needs and complexities.

"Our own Vanderbilt University Phase I resumption of specific, limited, on-campus operations and activities will begin May 18 and will be tailored to our own unique density, operations and other considerations as a residential education institution," Wente announced.[62]

Six days later, Wente and Eric Kopstain offered more insight into the university's thinking during a virtual town hall attended by more than 2,700 members of the Vanderbilt community. Interest in the university's plans was high; audience members submitted more than two hundred questions for the event.[63]

During the town hall, Wente and Kopstain outlined what life might look like if campus reopened, putting forward now-familiar elements of the Return to Campus Plan that they and their colleagues had been considering and discussing with board members. Reopening would support mission-critical activities. Planning would be data-driven, proactive, and based on guidance from local, state, and federal public health authorities (to the degree that such guidance was available) to keep the campus community as safe as possible. Campus resources, including physical space, would be deployed in "radically different ways." And the university would "plan for the best and prepare for the worst."[64]

Kopstain and Wente also talked through the baseline health protocols, including temperature monitoring and testing, that would be established for everyone on campus. Foot traffic in indoor and outdoor spaces would be designated "one way," they noted, to help ensure physical distancing. Directional signs would help manage circulation of people, restrict building occupancy, and remind people to follow protocols. University mental health and wellness resources, they noted, would be promoted.[65]

"We must embrace shared responsibility and be accountable for our actions at an individual level," a slide from Wente and Kopstain's presentation read. Another emphasized the importance of responding as "One Vanderbilt."

The administrators noted that bringing students back to campus would require the university to do things it had never done, like operating a testing and contact tracing program. New housing density limits would have to be imposed. Student health-care resources would need to accommodate any students who fell ill with COVID-19. Adjustments to the university calendar could mean unusual start and end dates for the semester, different breaks, or phasing in of various degree programs. Class schedules might have to shift to permit more time between classes. Reopening would require adopting best practices for "hybrid" instruction, with some students present in class and others participating remotely.[66]

For all the details they offered, Wente and Kopstain stopped short of saying definitively that Vanderbilt would reopen. A number of universities had made that announcement as far back as April,[67] but Vanderbilt's leadership wanted more time to make sure they could get reopening right. At a May meeting of the Board of Trust, Vanessa Beasley said that, among other factors, administrators were considering local public health guidelines and decisions made by peer universities. Also, given the unpredictable trajectory of the pandemic and all that was still unknown about the virus, the university wanted to preserve "optionality," Chancellor Diermeier says.

"We were clear that the world would change. So you want to have a lot of flexibility, and you want to make decisions as late as you can. You want to make them when you have to make them, but not earlier," he says. "The value in that is you learn more [in the meantime], and that would make [reopening] easier."

As promised, the university launched Phase I of its Return to Campus Plan—the "Prepare" phase—on May 18, five days after the virtual town hall. Most significantly, the phase allowed for research at more than one-third capacity, subject to a minimum of six feet of physical distancing and campus-wide safety protocols, including masking. Field training for graduate and professional students was allowed as informed by research activities or authorization from clinics and other occupational sites. Staff could be recalled to campus when needed to directly support research, instruction, and residential living. No gatherings were permitted.[68]

Resuming a limited degree of in-person research was, in some ways, a test case for wider reopening, Kopstain had noted during the May 13 town hall. "This will be an important opportunity to apply campus and building protocols more widely in a relatively discrete set of spaces," Kopstain said in the discussion. "It's very important that we get this right as a foundation for further scale-up of campus activities."[69]

Fully reopening Vanderbilt's in-person research operations was no less a priority than bringing students back to campus. By the time the campus had been all but closed down in March, faculty had submitted even more research proposals than they had at the same point the year prior.[70] But with the exception of COVID-19–related collaborations among university and VUMC faculty,[71] all on-campus labs had been restricted to "essential activities" on March 16; some research operations had continued remotely.

Says Cornelius Vanderbilt Professor of Molecular Physiology and Biophysics Alyssa Hasty: "Our biggest concern in Basic

Sciences was how to keep labs going and how to be equitable in that. For example, we have animals we have to take care of, and we have to give them special diets twice a week in order to follow our animal care protocols. So we had to find a way to choose who did that work and set up a system so that it could be done safely."

Researchers' federal funding agencies, thankfully, had allowed faculty, staff, students, and postdoctoral students to continue to be paid during the shutdown, regardless of the nature of their work.[72]

In the spring, Susan R. Wente had charged Vice Provost for Research Padma Raghavan, along with the deans for research from across the university's schools and colleges, to begin outlining the process. The team structured a phased approach for slowly opening labs and prioritizing faculty programs that "would be delayed significantly to the point of irreparable damage if not restarted," along with "research that could be accelerated for breakthroughs" in emerging topics. Core facilities and resources would be made available if they benefited multiple researchers and if they could be operated at less than peak capacity with additional protections, were "naturally COVID-19 protected," or were key to accelerating breakthroughs in emerging topics.[73]

Researchers began returning with the launch of the Return to Campus Plan on May 18. By May 22, sixteen on-campus buildings and four off campus had undergone a rigorous certification process that allowed them to reopen. The checklist included an inspection of all mechanical, electrical, and plumbing systems; review and necessary adjustments of ventilation systems; a clear circulation plan to direct foot traffic; delivery of an ample supply of cloth face coverings; and placement of signs indicating proper distancing and reminding researchers of other required protocols. Capacity for research would increase again when the university moved to Phase II of the Return to Campus Plan on June 8.

Lab personnel crafted solutions for resuming their work despite limited capacity. For example, members of Assistant Professor of Chemical and Biomolecular Engineering Ethan Lippmann's research group returned in staggered shifts to their lab in Olin Hall, the only place where they could restart cell cultures and analyze samples. When capacity would later expand to 50 percent during Phase II+ of the research ramp-up, the team would organize shifts so that every member could come in for half a day every day.[74]

With the onset of COVID-19, some faculty had shifted their research to focus on the disease. Biological physicist John Wikswo and his team were developing ways to get vaccines to market faster. Chemist David Wright and his colleagues were studying people who'd recovered from COVID-19 in an effort to invent faster tests for SARS-CoV-2. Mathematician Glenn Webb was building models to predict the size of the COVID-19 outbreak and studying the effect of interventions like social distancing. And anthropology professor and global public health expert T. S. Harvey collaborated with chemistry research professor and mobile health expert Thomas Scherr to develop an app enabling individuals to assess their risk of being infected with COVID-19. The app also offered capabilities for expediting test screening for providers and ultimately giving public health officials real-time anonymized data to identify, map, and target interventions where they were needed most.[75]

Vaccine research at VUMC, progressing at a pace once thought impossible, had gotten a high-profile boost in April, when pioneering music legend Dolly Parton, an established supporter of the medical center, donated $1 million to support the effort, drawing international headlines. Parton made the gift in honor of her longtime friend Vanderbilt surgery professor Dr. Naji Abumrad, whom she met in 2013 after being involved in a car accident. Her support helped fund development of mRNA-1273, the promising vaccine Mark Denison's lab was testing with Moderna.

In May there was more extraordinary news: Denison and his team had analyzed the blood serum of the first humans to be injected with mRNA-1273 and discovered that the vaccine wasn't just working, it was obliterating the virus. Denison phoned former Vanderbilt researcher Dr. Barney Graham, whose work at the National Institutes of Health had been key in the vaccine's development, to share the news. "We were the very first humans to see the effect of this vaccine actually killing COVID," Denison says. "It was a very emotional moment."[76]

The same month that Parton made her gift, VUMC researchers led by James Crowe and Robert Carnahan, who had raced since the pandemic's earliest days to identify antibodies that could be used as a COVID-19 therapy, had joined the global biopharmaceutical company AstraZeneca in the effort to pinpoint candidates for antibody-based treatments. By late spring, they had identified antibodies capable of neutralizing the virus and were rapidly selecting the best candidates for clinical development.[77]

There were other important projects underway. VUMC had taken on a key role in a national effort to establish a registry of US health-care workers and test whether hydroxychloroquine was protecting them, their patients, and their families from

COVID-19.[78] Other researchers at the center were recruiting volunteers for the first phase of a study of how blood plasma from recovered patients—a strategy used for more than a century to treat a wide variety of infections—might be used for treating COVID-19.[79]

Vanderbilt's leadership knew that compliance with safety protocols by a majority of the university's students, faculty, and staff would be essential to containing the virus if campus reopened in the fall. Regularly reminding people of protocols and expectations would be crucial—and so would winning the cooperation of students, especially. Staff in the Division of Communications had begun discussing development of a "hearts and minds" campaign as early as April.

"We quickly realized that if we were going to bring students back to campus, this was going to be an exercise in behavior change," says Daniel Dubois, executive director of marketing solutions. "And to actually influence behavior, you can't just wallpaper the campus with protocol signage. You need a well-crafted, thorough, cohesive, and long-term campaign."

In his previous job with World Wildlife Fund, Vice Chancellor for Communications Steve Ertel had run campaigns aimed at getting corporations on the right side of climate change and other urgent environmental issues. He saw similarities between those campaigns and the one Vanderbilt needed.

"In both cases, our job is to inform and inspire—to make sure people understand what they need to do and inspire them to want to do the right thing. And for those who don't, we also need to communicate that there are consequences," he says.

Vanderbilt's campaign "was about inspiring compliance," Ertel explains, "because the only way bringing students back is going to work is if people get their tests, wear masks, socially distance, and avoid large gatherings."

First, Ertel's team tapped Dr. Gerald Hickson of VUMC's Center for Patient and Professional Advocacy. Before the pandemic, Hickson had conducted extensive research on persuading more physicians to comply with handwashing guidelines aimed at stopping the spread of pathogens.

"His research received widespread acclaim because he had so much great data about the right mix of sticks and carrots, and how to actually change behavior," Dubois says. "Another thing Dr. Hickson was very clear on, which helped us a lot, is the importance of humor. You're not going to win anyone over just by bludgeoning them with rules."

The team also consulted Kelly Goldsmith, a behavioral scientist and associate professor of marketing at the Owen Graduate School of Management. Goldsmith's most recent research had focused on scarcity and consumer patterns since the start of the pandemic.

Ertel's team reviewed a range of anti-smoking and anti-drug campaigns, as well as campaigns promoting condoms and safer sex. Though different in their approaches, all of those campaigns ultimately had the same goal: getting someone to do something they don't want to do.

At first, "we had a lot going against us," Dubois acknowledges—like the pandemic's many unknowns and the fact that most Vanderbilt students were in an age group that appeared to be less likely to suffer dire effects from COVID-19. "There was a sense among students of, 'We're highly social, we're immortal, you can't tell us what to do,'" he says.

He and his team aimed the campaign at students predisposed to doing the right thing, those who believed they could play a role in keeping others safe. "The objective was to give them something to rally around," Ertel says.

Throughout May, drawing on their research, the team developed and tested possible campaign themes. After considering several options, the team concluded those approaches didn't quite convey either Vanderbilt's values or the need for accountability. One weekend at home, Dubois enlisted his children in a brainstorming session. Among the words they wrote on the whiteboard in the family's basement were "Anchor Down," the university's popular, all-purpose rallying cry. The team built on that, arriving at "Anchor Down. Mask Up." It was good, but not broad enough to encompass all the safety protocols—like handwashing and physical distancing—that the campaign needed to promote.

On Memorial Day weekend, after hiking in the Fiery Gizzard recreation area southeast of Nashville, Dubois checked his email in the parking lot before driving home. He found a message from Ertel, suggesting a campaign theme: "Anchor Down. Step Up." The slogan's emotional connection to Vanderbilt was clear. So was its call for personal action and responsibility. And the line could serve as an umbrella slogan for four additional messages promoting the main safety protocols: Mask Up, Wash Up, Back Up (for physical distancing), and Check Up (for symptom monitoring). Twenty-four hours later, Dubois and his team had mocked up logos, and Ertel had written the script for an inspirational "anthem video" that would be the centerpiece of the campaign.

Another Crisis Rears Its Head

The same weekend Ertel and Dubois had their breakthrough, a clerk at a Minneapolis convenience store called the police to report that a man had paid for cigarettes with a counterfeit twenty-dollar bill. What would happen next cost the man his life, defined the summer, further exposed the unhealed wounds of a nation, and demanded that Vanderbilt's leaders contend with another challenge. In addition to public health and financial crises, Vanderbilt and the nation now faced a moral crisis.

The killing of George Floyd was the latest in a seemingly endless stream of deaths of Black Americans at the hands of police. After the delay in making arrests for the killing of Ahmaud Arbery in Georgia in February and the fatal shooting of Breonna Taylor by Louisville, Kentucky, police in March, Floyd's death sparked grief and rage and renewed calls for police reform and racial justice across the country. Tens of thousands marched in protest in cities nationwide—including Nashville—in the days after Floyd's death. Protests in Minneapolis and elsewhere turned violent.

Vanderbilt, like many universities, had in recent years begun grappling with the racism intrinsic in its past and present. Four days after George Floyd's death, the university issued a short statement that said, in part, "In all our roles as educators, scholars and stewards of the best ideas and values of a civil society, we must be part of the national chorus that cries out that all lives are valuable. Furthermore, as a community we should question why incidents such as these continue to occur, and what role we can play in creating a fairer and just world."[80]

The university also accelerated changes in its leadership that were already in motion. On May 27, Interim Vice Chancellor for Equity, Diversity and Inclusion Dr. André Churchwell was named chief diversity officer. Several other interim appointments were made official, including that of William Robinson, who was named permanent executive director of the Office of Inclusive Excellence under the auspices of the provost's office. Daniel Diermeier had asked that the interim positions be made permanent so that the university could move forward quickly on issues of diversity, equity, and inclusion as soon as he assumed the chancellorship July 1.[81]

Saturday, May 30, brought more peaceful protests—some of the largest seen in America in generations. But as night fell, demonstrations in some cities, Nashville included, again turned violent. Americans, already anxious and weary from the pandemic and traumatized by the footage of George Floyd dying on the street beneath a policeman's knee, watched as buildings were set ablaze and protesters and police clashed. The next day, Susan R. Wente issued a statement that, like the first, reiterated Vanderbilt's rejection of "racism, prejudice, hatred or violence in any form."[82]

"[W]e will not be silent, and we must address the root causes that have driven us to this point as a society. The open wounds of racial injustice and inequality and structural barriers to equity have festered in America for too long," she wrote.

The same day, in a private email to Churchwell and Robinson and a few other administrators, Wente wrote, "We are yet again called to make our voices heard and to take actions that will heal our community but also make our community a model of inclusion, equity, dignity and belonging."

Churchwell published his own statement the same day.

"My heart is broken," wrote Churchwell, a cardiologist. "... [M]ore than one virus infects this country; the older virus is racism . . . If we view our city and country as a 'broken heart,' we can, with the correct 'treatments,' guided by our faith, morals and beliefs in the infinite possibilities of a single life, mend our broken heart. Indeed, we MUST.

"We must also believe that our great institutions, like Vanderbilt, can be part of the treatment and new solutions to help our city and country heal."

In the days that followed, the university held several online listening sessions—sooner than originally planned, at Wente's urging—for students, staff, faculty, and postdoctoral scholars. Churchwell, Kopstain, Senior Associate Vice Chancellor for Public Safety and Special Initiatives and Chief of Police August Washington, and other leaders hosted the sessions.

In early June, Diermeier hosted a series of "virtual gatherings," livestreamed on Zoom and YouTube, with students and their families, faculty, staff, and alumni. The main objective was for the incoming chancellor to introduce himself, share his vision for Vanderbilt, and offer reassurance and empathy amid the epidemic. But he also answered selected questions that viewers submitted in advance, including questions about race and equity issues.

"We've done a lot at Vanderbilt in the area of diversity, equity, and inclusion," Diermeier said in the gathering for faculty. "But we all have to ask ourselves, 'How do we do more in combating racism and social injustice?'"[83] He suggested that Vanderbilt "double down" on recruiting Black students and faculty.

In his comments to students and their families, Diermeier reflected on the role of the university during the pandemic and the nation's reckoning with police brutality and racism.

"It's in difficult times that universities can play an important role, and when their value is particularly obvious, because those are the places where we can look at the causes, the conditions, and the complexities of the various problems that we have to deal with in society," Diermeier said. "It's the place where we can . . . discuss, debate, and learn, where we have a community of students and scholars that can figure out solutions to these challenges and make a lasting contribution and improve society and mankind more generally."[84]

On June 29, hundreds took part in the university's virtual memorial service for Black lives lost to police brutality and racism, which was livestreamed on YouTube from Kirkland Hall and hosted by Churchwell, who had suggested the event because, he recalled later, "Our community needed to see we were grieving, as a team." During the service, Wente officially apologized to the "cherished members of our Black community across generations" for "the pain inflicted" by the university's racism and past lack of action.

"We cannot move forward as an inclusive university dedicated to bettering mankind without acknowledging our own racism and taking actions," Wente said.[85]

In remarks during the service, Diermeier said, "We know that this memorial must be followed by action . . . The answers are not quick, and they are not easy, but finding them and mapping our next steps is urgent, and we will begin this work today."[86]

When Diermeier officially became Vanderbilt's ninth chancellor on July 1, he sent two emails to the Vanderbilt community. One was a typical first-day message, outlining his goals and vision. The other—co-signed by Wente and Churchwell—described in more than 1,100 words how the university planned to address racial injustice. It proposed expanding efforts to diversify the cohorts of faculty, staff, and students. It called for "significantly increasing" the budget for the Office for Equity, Diversity and

Inclusion and for bolstering bias and inclusivity training, partnering with the city of Nashville, creating an ad hoc Board of Trust committee, and confronting the racism in Vanderbilt's history, among other actions.[87] The message was the result of a "collaborative, interpersonal set of discussions among leadership, including all the vice chancellors, in terms of how we would approach this," Churchwell says. "No controversy, no turbulence. Just a single-minded approach to doing the right thing."

Since being named chancellor, Diermeier had spoken often about the importance of balancing short-term priorities with longer-term goals. His day one message offered reassurance that the university would not lose sight of fighting inequity amid the urgent and seemingly endless demands of the COVID-19 pandemic.

The university's resolve was tested just two days later when an Instagram video of a Vanderbilt student using a racial slur renewed debate about racism in the university's Greek system.[88] In a statement, the university said it "condemns the language in the strongest possible terms," had referred the matter to its Title IX and Student Discrimination office, and was investigating the incident. "It is critically important that our community understands that this type of language, which undermines and erodes the culture of belonging and safety we have worked to create, is unacceptable in any context," the statement read.[89]

Daniel Diermeier's assumption of the chancellor role marked the quiet end of the months-long and mostly seamless process in which Susan R. Wente had handed him the baton as Vanderbilt's leader. In the eleven months since she had taken over for Nicholas Zeppos, Wente had steered Vanderbilt through what was arguably the most harrowing and chaotic period in its history.

"The problem-solving was continuous. And it was reiterative," Wente says of the experience. "I'm a research scientist. I need data. This was planning for multiple scenarios with limited information, and that's exactly what you do when you're planning an experiment. So I felt very comfortable in that kind of mindset. I felt comfortable in the fact that there were unknowns."

Through it all, she evangelized for the idea of One Vanderbilt and "the Vanderbilt way" of cross-departmental collaboration, championing what she called the three T's: trust, transparency, and teamwork. Before the pandemic, the principles had been Wente's refrain in everything from an annual address to faculty to her biweekly column on Vanderbilt's website. For her and for the university, they continued to serve as guiding values through the challenges of 2020.

A Decision Is Made

By the end of May, the number of COVID-19 deaths in the United States had passed the 100,000 mark,[90] and the pandemic showed no signs of slowing as summer approached. By June, with COVID-19 cases on the rise in nations where the number of infections had initially been relatively low, the number of cases worldwide had reached nearly seven million and deaths surged past 400,000.[91] Even as strict lockdowns began to ease in New York and London, cases were spiking in the southern and western US, including states where governors had eased restrictions earlier and had not imposed mask mandates.[92] In late June, thirty-six states reported increasing cases.[93]

However, in Nashville, new cases of COVID-19 had dropped a bit since their peak in May.[94] According to a presentation by university leadership at a June Board of Trust meeting, the transmission rate in the city was at 1—relatively low, although still above the city's goal rate of less than 1. The city had been in Phase 2 of its reopening plan since May 25, allowing restaurants and some businesses to reopen at 75 percent capacity and salons and gyms to open at 50 percent.[95] By mid-June, Vanderbilt's leadership decided the time was right to announce its plan to reopen.

Administrators presented the plan to the Board of Trust in June. In addition to the elements they had outlined since April, the plan called for holding in-person classes for the fall semester with a "fully developed contingency plan" for online teaching that could be activated as needed. All classes would have options for online delivery for students who could not come to campus because of travel restrictions or health risks, and for any students on-campus required to quarantine or isolate. Additional resources would be invested in online instruction to "allow for the very best educational experience" and provide for faculty in high-risk health groups who could not teach in person. Travel breaks would be eliminated, reducing the risk of students, faculty, and staff spreading the virus elsewhere or bringing it back to campus. In-person classes would end in November, and students would return home before an expected uptick in COVID-19 cases in the late fall and early winter.

On June 16, the university made its plan public.

"We successfully completed Phase 1 with the initial ramp-up of our on-campus research activities, and on June 8 began transitioning to Phase 2, with a further expansion of research and graduate and professional student experiential learning," Diermeier and Wente said in a joint announcement. "All of our planned phases are grounded in the adoption of rigorous safety and prevention protocols. Based on the effectiveness of these actions, and after studying multiple scenarios and best practices, we determined and are very pleased to announce that we will resume on-campus, in-person classes for the fall semester."[96]

Classes would begin on August 24 and conclude on November 20, just before Thanksgiving, with final exams for undergraduates administered online. All students were invited to return to campus for the fall semester, but the university promised an equivalent, equally robust experience for students opting to receive instruction remotely. In any case, classes would "look and operate differently from the way they have in the past," Diermeier and Wente said, to accommodate physical distancing. Students' course schedules might include evening and weekend classes, a blend of virtual and in-person learning, and other approaches. COVID-19 testing for students would be mandatory, and the university was committed to ensuring that all students could be cared for if they were ill or needed to quarantine.[97]

After months of planning and almost nonstop work, Vanderbilt was officially on the road to reopening.

"We were a little later than most in announcing, but that was by design," Diermeier says. "We really wanted to make sure that we had 80, 90 percent confidence that we could pull this off, that we could see a path. That's when we made the announcement."

Not everyone welcomed the news. Skeptics of reopening any college campus during the pandemic—"the doubters," as Diermeier refers to them—were plentiful.

"Whether colleges are willing to admit it or not, chaos will be greeting many of them in the coming weeks, and wishful thinking will not be enough to avoid it," wrote Robert Kelchen, an associate professor of higher education at Seton Hall, in the *Chronicle of Higher Education*.[98]

"Gathering students on campus is a gamble that could generate outsize risks for society and only modest benefits for students," argued an economist in *The New York Times*.[99]

Writing in *Inside Higher Ed*, Janet Murray, a professor at the Georgia Institute of Technology, likened the reasoning driving reopening to the engineering groupthink that resulted in the fatal failure of the O-ring on the Space Shuttle *Challenger*.[100]

Beyond the pundits, there was a crucial Vanderbilt constituency that was not uniformly in favor of in-person classes: the university's 1,200 or so faculty members. The people who would be working in close proximity with students had, since reopening discussions began, voiced serious reservations. Vice Provost for Faculty Affairs Tracey George described faculty responses as "variable, but largely negative and highly critical, with certain themes coming through loudly: uncertainty, suspicion, and anger. Very few were positive about the decision. Even those who were positive questioned the soundness of our reasoning, given approaches taken by our peers."

Indeed, Vanderbilt was in exclusive company. Around the country, universities' plans for the fall had expanded and contracted for months, with some schools announcing plans for in-person teaching as early as spring, only to walk them back when COVID-19 case numbers made the plans untenable.

By late summer, Vanderbilt was one of a small number of the nation's top-ranked universities inviting all of its students back to campus and aiming to teach as many classes in person as possible.[101]

On July 24, the Vanderbilt chapter of the American Association of University Professors sent Diermeier and Wente a petition calling for online instruction, rather than in-person instruction, to be the university's default option. The petition was signed by about two hundred faculty members and several hundred more students, parents, and alumni. "Current plans to resume extensive in-person teaching and campus activities will almost certainly result in serious illness and even death for some in the Vanderbilt community," said a guest op-ed from the AAUP in *The Vanderbilt Hustler*. It noted that "peer institutions that also prioritize residential education, including Harvard, Stanford, Johns Hopkins, Columbia, the University of Pennsylvania, and Princeton, have all switched to total or near-total remote teaching in the fall."[102]

Diermeier was empathetic to instructors' concerns but resisted issuing a university-wide decree dictating what they would do or not do. Instead he took a decentralized approach, where deans and the department heads would address faculty concerns on a case-by-case basis. "I was confident enough in our collaborative culture that we would work this through. We make these decisions on teaching always school by school, college by college, department by department, and I said, 'That's the way we're going to do it. We're going to work through things together.'"

Four days after the AAUP's op-ed, a group of fifty or so parents and students sent an open letter to Vanderbilt's leadership, citing "grave concerns" about aspects of the Return to Campus Plan. The group asked that the university's surveillance testing program be strengthened, and they articulated concerns about the limits of contact tracing and self-reporting, the possibility of airborne virus spread through HVAC systems, and more.[103]

Even staff weren't sure about reopening when the plan was initially announced. "There was a lot of doubt," Mark Petty says. "I don't think it wasn't peppered with optimism, but it was just, 'How do you make this work?'" But by mid-July, he adds, he was convinced. By then, more measures were in place, and he understood that the fast-changing pandemic might look better for Nashville by mid-August.

Looking back, Wente says she believes many objections to the university's plans were about something bigger.

"People were scared and upset at all the uncertainty and disruption in their lives. They had to get angry at someone. You couldn't get angry at the virus—the virus was something you had no control over. We took a lot of heat for any decision we made. Everybody felt like we were treating somebody badly—staff, faculty, or students. Because there was nothing you could do with your anger about the situation . . . and feeling like you had no control over your life. So we became the target. And we just had to keep calm, keep explaining, keep providing rationale. We had to keep saying we would pivot and change if we needed to."

Some critics' fears appeared to be validated when COVID-19 outbreaks were reported at several fraternities around the country over the summer, and when a number of universities reversed plans to offer in-person classes.[104]

Vanderbilt's administrators responded to concerns as they arose, but, as the August start date drew nearer on the horizon, they stuck to their plan and to their contention that the safest place to be in Nashville come August would be on the Vanderbilt campus, thanks to safety protocols designed especially for the campus community.

"We asked ourselves, are we providing as safe an environment as people would have at home? And the answer was yes," Diermeier says. On campus, students and faculty would have access to the surveillance testing and contact tracing that were missing in most American communities, and one of the world's best medical centers would be next door. "So when you come here, you're actually better off than at home," Diermeier says. "And we said, 'But if that is still not enough for you, you can attend classes remotely.' It was important for us that we give students the choice."

Vanderbilt stayed its course. However, there was still much to do, from finalizing the testing program, to sorting out teaching and housing, to creating adequate space for physical distancing—and just weeks left to do it.

A New Kind of Campus Life

In the spring and summer, as information on the university's COVID-19 website continued to grow, Vanderbilt's communications team saw the need for a better system for conveying the details to the university's varied audiences. Their solution was to create a series of comprehensive FAQs tagged for each audience.

On the day that plans for the fall semester were announced, Executive Director of Digital Strategies Lacy Paschal and her team officially transformed the COVID-19 site into the "Return to Campus" website; it had more than three hundred FAQs.

"We want parents, students, faculty, and staff to have the information that they need," Paschal says. "I have a first-grader, and I read his school's plans for their students' return fifty times. So I get it. We want people to find the information they're looking for and not get frustrated."

The aesthetics of the site also shifted. For months, vanderbilt.edu/coronavirus had featured a bright blue header with renderings of ominous viruses. Now, there was a purposeful change to a more hopeful tone—conveyed by a Commodore gold color palette and photos of campus.

During the site's development, there were few similar sites to look to for inspiration. "So few schools were going back that we were in a unique situation," Paschal says. "In fact, we had people from other institutions coming to us, asking 'How are you structuring this content? How does this work?' I had conference calls with my counterparts at other universities who wanted to know more about the FAQ model and how we had managed to centralize everything."

Included among the comprehensive list of FAQs on the site were details on testing and tracing, dining, housing, and other questions every parent and student no doubt had.

Testing and Tracing

Pam Jones, a senior associate dean in the Vanderbilt School of Nursing, had been thinking since March about what the university's COVID-19 testing program might look like; she'd been asked to help convert campus health clinics to testing sites shortly after the pandemic hit. Susan R. Wente had also tapped her to join the university's Public Health Advisory Task Force.[105] Now Jones and Andrea George, Vanderbilt's director of environmental health and safety, were co-commanders of the Public Health Central Command Center, or PHCCC, the organization charged with creating a testing and contact tracing program for students' return in the fall. Among the big questions Jones and George had to answer: Which test was best? Who would be tested? Where? And how often?

A draft plan that had taken shape by May called for the university to partner with VUMC to employ the medical center's best testing practices, including use of polymerase chain reaction (PCR) testing using nasal or nasopharyngeal swabs. That test was considered the gold standard for accuracy. VUMC had been able to process seven hundred to eight hundred of the tests per day, with a twenty-four- to forty-eight-hour turnaround. If the FDA approved a saliva PCR test or rapid turnaround antigen test with high efficacy rates between then and the fall, the draft plan noted, VUMC and the university would switch to that. Vanderbilt had considered establishing a new testing system with researchers in the School of Medicine but determined it would be more expedient to work with a third-party vendor that had IT and HIPAA-compliant procedures already in place.

Over the summer, George oversaw development of the testing program, while Jones was charged with working out contact tracing. An environmental engineer by training,[106] George recalls the necessity of staying on top of the news and the constantly evolving local guidance and medical information. "Many of us would get up at 4 a.m. and just start reading whatever had been published the day before. We were also trying to see what our peers were doing, while leaning on guidance from our colleagues at VUMC," she says.

George and her team soon determined that undergraduate students—because they would be living in congregate settings—would be required to test at home within two weeks before arriving on campus. If they tested positive, they were expected to isolate at home and await medical clearance. Once they arrived, all students would be screened and tested in a large-scale operation housed in the David Williams II Recreation and Wellness Center. Anyone testing positive would isolate, and their recent close contacts would be notified.

Making decisions in the rapidly evolving environment of the pandemic could be challenging. "As dedicated academics, it is in our nature to want to gather as many opinions and as much information as possible before moving forward," George says.

Jones agrees. "I have never in my career—and I've been at this a long time—faced anything where we knew so little. Other disease processes follow a relatively predictable course, but this doesn't. You don't know if you're going to be fine and not even feel a symptom, or if you're going to die. And that's a pretty scary thing. And we've got CDC guidelines that keep changing. So you just make the soundest decisions you can, implement them well, and hope for the best."

There were bumps in the road. While putting the pieces in place to test all students upon their arrival in mid-August, the PHCCC team discovered that the thousands of testing kits they had ordered were not pre-assembled. "We got the kits in pieces, not pre-packaged and ready to go like we'd expected," George recalls. "So we had all the pieces shipped to the rec center, and we put out a call for volunteers to participate in a socially distanced test kit creation day." In true Vanderbilt style, plenty of helpers showed up.

By the first of August, the command center was running at full capacity. From June 29 to August 4, the university tracked nearly seven thousand COVID-19 test results. In a community-wide "Return to Campus" email on August 10, Vanderbilt announced that there were twenty-seven positive cases within that pool, none of which, according to contact tracing, were the result of on-campus contact. This promising sample pointed to the strength of the protective measures the university had put in place over the summer and would expand in the run-up to fall.

All undergraduates were mailed test kits and required to submit samples before returning to campus; they would be tested again immediately upon their arrival. The university continued to adjust its testing strategy according to the emerging scientific understanding of best practices. The August 10 email had announced that there would be regular testing of "a sampling of asymptomatic community members based on analysis of COVID-19 test outcomes." In the end, though, the university judged that keeping campus as safe as possible would require weekly testing for all undergraduate students not enrolled in remote study.[107]

In the development of the testing and tracing program, and throughout reopening, Vanderbilt's School of Nursing was an invaluable resource. The school's expertise informed numerous practices and protocols, as well as on-campus messaging. As campus pioneers in online instruction, they had been able to pivot quickly in the spring and provided advice to faculty members across campus who were preparing for the fall. And at VUMC, nurse practitioners who'd trained at the School of Nursing led the care of gravely ill patients in the COVID-19 intensive care unit.

"Nurses are trained in the discipline of community health and public health," Jones notes. "This is our sweet spot."

De-densification

Another critical element of the Return to Campus Plan—and one of the greatest challenges it posed—was "de-densifying," the awkward term that had come to be used during the pandemic to describe providing enough space to allow for proper physical distancing. Green spaces aside, a university campus is designed to be dense. It's meant to accommodate classes filled with students, researchers laboring together in labs, close conversations during seminars, communal meals, collaborative office work, music rehearsals, and athletic practices and contests. The point of a university campus is to bring people together.

As Eric Kopstain puts it, "We quickly realized that the whole physical environment of our campus would have to be different."

Shortly after students left in March, Megan Sargent, a structural engineer and project manager in campus planning, started thinking about the flow of people through Vanderbilt's buildings and grounds.

"It felt pretty abstract at first," says Sargent, who worked closely on the project with a GIS analyst and landscape architect. "Like everybody, we were starting from scratch. I remember googling how people were approaching circulation solutions in Singapore. It was daunting, but it was also exciting."

Sargent met regularly with representatives from the university's mobility, dining, sustainability, and emergency preparedness teams to consider the range of distancing needs and their implications. In the spring, her team started the rigorous process of mapping each building and routing foot traffic. Although they started with certain universal principles—including the ideas that people should only move clockwise through buildings and that a door should either be an entry or an exit, but never both—it quickly became clear that some on-paper ideals would not always translate into physical reality. For example, many of the campus's older buildings only had one point of entry accessible for someone in a wheelchair. There were also university ID card access points to consider.

Dining halls quickly emerged as circulation "pinch points," because rushes of students tended to fill them at specific times every day. And despite strict maximum capacity signs on the entrances to bathrooms, the number of occupants would be hard to enforce or even gauge; it wasn't really possible to count people in stalls.

Then there was the Stevenson Center, a notoriously hard-to-navigate cluster of academic buildings dedicated to science, engineering, and math. "It felt as though, with every iteration of our plans for the center, a pathway would work for one person but it wouldn't work for others," Sargent recalls.

Sargent says that, overall, some of the team's conversations became pretty philosophical.

"We wanted to guide people and take away some of the guesswork for them, but without being overly prescriptive," she says. "There were a lot of times when we asked ourselves, 'Are we going too far?' After all, we wanted to respect the fact that there are already so many stressors in people's lives. We wanted them to come outside and enjoy the campus spaces and not just feel like everything is off limits."

Classrooms received especially careful attention. "We were super worried about transmissions" there, Diermeier acknowledges. While the tasks of creating distanced configurations and determining new maximum capacities belonged to individual building managers, the university also assembled a team of several vice provosts and other campus leaders to advise on best practices. The team had to strike a delicate balance. As Vanessa Beasley put it in a July 2020 virtual town hall for faculty, there was a need to "respect faculty's decisions and personal pedagogical styles while also making sure every class is as safe as possible for everyone who enters."[108]

Students would be prohibited from eating and drinking in classrooms and were required to wear a mask at all times. Faculty would receive their own lapel microphones so they wouldn't have to share handheld mics with colleagues. In lecture halls, many chairs were marked "off limits" to keep students from sitting too close to one another. Building managers worked with the circulation team to ensure at least six feet of separation between the instructor and the students.

Following guidance from the CDC and the American Society of Heating, Refrigerating and Air Conditioning Engineers, facilities teams increased the amount of outside air delivered to classrooms and other spaces and deployed high-efficiency filtration systems, along with new technology for monitoring temperature and humidity. To prevent the spread of the virus on high-touch surfaces, classrooms would be sanitized twice daily.

Bathrooms were also high on the custodial team's list, given the broader daily use of sink handles, towel dispensers, and other equipment. In the lead-up to the fall, facilities staff gradually replaced hand-operated machines such as soap dispensers with "touchless" alternatives wherever possible and installed foot pulls on the doors of restrooms and other high-use areas.

Vanderbilt's 340 acres of outdoor space,[109] paired with Nashville's relatively temperate climate, were advantages when it came to spreading out. Just as building crews had put down blue floor tape to delineate traffic pathways inside buildings, the turf and facilities team spray-painted hundreds of circles on Vanderbilt's lawns to demarcate social configurations allowing for six feet of separation. "There was a team that painted circles on just about every available green space," says Sports Turf and

Facilities Manager Ryan Storey. "I spent some time myself bent over with a tape measure and cans of paint. I was walking backwards so much to draw the circles that I made myself dizzy."

Storey and his coworkers also placed directional stickers on the ground to guide traffic and hung many exterior signs and banners promoting the effort to contain the virus. They also helped set up three large tents on Alumni Lawn, Library Lawn, and Peabody Esplanade—complete with radiant-heat flooring—that would be used for physically distanced dining, classes, exercise, and other activities once students arrived.

Housing

Reimagining student housing was a particular challenge of de-densification. One of Vanderbilt's highest priorities, and, along with resuming research, one of the two main reasons it was reopening campus, was to create a close-knit community that lived, ate, learned, and played together. Sharing a dorm room with a roommate or two was part of that residential education and also a rite of passage for college students everywhere. But would any of it be possible during the pandemic?

The university's June 16 announcement of its fall plans had identified "three cohorts of students with different [housing] needs," including first-year students, returning upper-division students, and student-athletes. Housing for student-athletes was relatively easy to solve; they were allowed to live alongside their teammates in acknowledgment of their inevitable close proximity during team practices and whatever games or travel might happen. Upper-division students were also permitted to live with roommates, since they lived in suites, with one student per bedroom. Says student Minna Apostolova, "At the time we were choosing who to live with, we had no idea that our living arrangements would be so important to our health."

First-year students presented different challenges. On one hand, administrators had concluded that all first-year students—who in typical years were housed in double-occupancy rooms—would, for safety's sake, need singles instead. At the same time, they would need some semblance of the traditional, formative initiation to life at Vanderbilt that came from living in The Martha Rivers Ingram Commons, which was defined by the very spirit of camaraderie and community that would be hard to foster within the limits dictated by safety protocols.

Described by Randy Tarkington as "one of the hallmarks of a Vanderbilt residential education," The Ingram Commons houses first-year students together in ten residence halls or "houses" alongside residential faculty members and peer mentors.

"The experience that students get can be life-changing. It's a tremendous way to start your college experience, so giving students a chance to have that was essential," he says.

The problem was that there weren't enough rooms in The Ingram Commons houses to provide single rooms for every incoming student, even after subtracting the students who had opted to learn remotely. The solution? A plan dubbed "the Flip," in which half the first-year students would live in the commons for the fall, and the other half would live in Carmichael Towers and Branscomb Quadrangle elsewhere on campus. At the end of the semester, the two groups would trade places. The goal of the Flip was to "create a whole first-year community, even though we're separated across two campuses," says Melissa Gresalfi, dean of The Ingram Commons.

Over the summer, as word of the Flip began to circulate, incoming students noticed that not all residence halls were created equal. In contrast to the brochure-worthy green spaces and colonial architecture of the houses that made up The Ingram Commons, Carmichael Towers consisted of two Brutalist structures built along West End Avenue in the late 1960s. Two of the towers had been demolished in 2019 to make way for new residential colleges. Another had been scheduled to come down in May, but the plan was halted once the university realized it would need the space to provide more single rooms. Some first-year students assigned to the Towers for the fall were lamenting both the distance that would be between them and their peers in The Ingram Commons, and the prospect of living in high-rise dorms that resembled a Soviet-era housing project. The Towers' reputation among students was less than sterling.

"The only thing on my mind was, 'not Towers,'" incoming first-year student Marissa Tessier told *The Vanderbilt Hustler*. "I opened the mail and thought, 'Why me?' I wanted to live in the commons."[110]

The housing reconfiguration forced one entire group of students off campus: those transferring from other colleges. The university usually offered its several hundred transfer students the opportunity to live on campus, but this year, there was simply not enough space to accommodate them.

As housing assignments were sorted out, Tarkington and his team were spending the summer "putting policies and procedures around everything." They had in-depth conversations about the expectations for residential advisers, who serve as counselors, social directors, and rule enforcers during normal times and who would be especially vital during a public health emergency. The team thought about how programming and community gatherings could translate to a virtual format. They discussed how to address students' mental health concerns—a big part of the responsibilities of the Office of Housing and Residential Experience—while also adhering to safety protocols.

Tarkington constantly tracked positivity rates for Nashville and the rest of the country and wondered if the semester would start as scheduled. "Everything felt like a moving target," he says.

He found a measure of reassurance in the information coming from Vanderbilt's leadership and the committees mapping out the reopening process. "They were making sound decisions," Tarkington says. "Everything was really well thought through. Plus, testing—and a lot of other things—were under discussion, which would give us the best possible chance to make it."

As the August 17 move-in date drew closer, it appeared that the de-densification of housing would be achieved in part by students simply choosing to live elsewhere. Less than 60 percent of undergraduates would be living on campus. Almost nine hundred were planning to take classes remotely. Of those who would be living on campus, more than 85 percent would have their own rooms.

Because housing now also needed to include rooms where undergraduates living on campus could quarantine and isolate as necessary, the university set aside spaces at several on-campus residence halls as well as a nearby retreat center. There was enough space to house about a quarter of the undergraduate population, if necessary. Undergrads living off campus, as well as graduate and professional students, would quarantine or isolate in their residences. Space in the Holiday Inn Vanderbilt would be available at steeply discounted rates. Vanderbilt biostatistics experts led by Professor Yu Shyr were developing a predictive model to further estimate quarantine and space needs, based on the same model being used to predict demand for VUMC's intensive-care unit.

Dining

In addition to being housed, all three thousand undergraduates on campus would also need to be fed. After scrambling in the spring to offer grab-and-go dining in consolidated facilities for the several hundred people remaining on campus, dining services teams were now readying to have students back on campus and aiming to provide some shade of communal dining while keeping everyone safe. It required them to reimagine the entire Vanderbilt dining system.

Students would have the option to dine in the tents of different sizes that had been raised around campus. The smallest tent would hold about 180 students at a time; the largest, about 280. Capacity in all would be metered to prevent overcrowding. Campus dining managers worked with the Design as an Immersive Vanderbilt Experience (DIVE) program at the Wond'ry to develop a series of tent protocols, including table sanitization, trash removal, and the metering system.

Also on the dining team's to-do list: revising cleaning procedures; developing new, streamlined menus; sourcing enough disposable containers and utensils to support a fully grab-and-go system; developing a network of pop-up dining locations; renting a fleet of golf carts to deliver supplies around campus; and installing plexiglass barriers for service areas. As was the case all around campus, signs and wayfinding would be crucial to helping students navigate the new system.

"Our team produced hundreds of informational and wayfinding signs," says Sean Carroll, Vanderbilt's director of marketing and communications for business services. "We also invested in digital menu screens placed at the entrance to queuing lines, to help students decide what they wanted to eat before they entered the line and allow for a faster experience."

As plans were implemented, the university's dining managers got little rest.

"From the day students were sent home, through the summer, our managers didn't take any vacation," says campus dining Executive Director David ter Kuile. "They were working on a rotating basis, but even when they were home, they were having to plan for when they came back—breakfast, lunch, and dinner. So really, they worked seven days a week, nonstop, all day, every day."

Teaching

Before the university announced its decision to bring students back, Vanderbilt's faculty had spent spring and early summer simultaneously preparing to teach online, as they'd found themselves doing in March, and to teach in person in radically altered spaces designed to accommodate physical distancing. With classes now officially scheduled to resume, but not every student willing or able to be on campus, faculty would have to quickly learn to teach "hybrid" courses, with some students present in class and others participating through Zoom. And there was always the possibility that classes could start in person but be forced online by an outbreak of the virus. It meant preparing for all eventualities.

As they began doing so, instructors' concerns ranged from the philosophical—like the costs to their research and scholarly work from spending so much time learning the demanding new mode—to the practical, like mastering remote-instruction technology and planning classes.

"How do we continue to provide a world-class education when faced with an unprecedented crisis?" Vice Provost for Faculty Affairs Tracey George had asked rhetorically at a Board of Trust COVID-19 committee meeting in May. Her answer: courses that were designed to be adaptable in their arrangement, timing, delivery, and accessibility. To get there, faculty would need thorough training, support, and infrastructure. Helping to provide all three were the university's IT department and Center for Teaching.

Over the summer, among other preparations, the Vanderbilt University IT team worked many long days to upgrade almost eighty classrooms with advanced technology and continued to put support systems in place for instructors. The Center for Teaching, meanwhile, was staffed by experts in instruction and education technology and instructional design. Even before COVID, the team had provided a range of support to the university's teaching faculty, from individual consultations to workshops to podcasts and books. Since the coronavirus had come to campus, the center's staff had been in high demand. They had interacted with nearly a third of faculty in 2019, but in the first six months of 2020, they were approaching 90 percent.

To help faculty expand their fluency in online teaching, the center was offering an online course design institute: intensive two-week online courses for as many as six faculty at a time. Mustering the capacity of support staff to facilitate the move to hybrid teaching was "the biggest pain point" in George's view, but with overtime and commitment, the work was getting done. Investments the university had made in the Center for Teaching and in classroom technology before 2020 proved crucial. Without them, George says, "fall hybrid teaching would simply have been impossible."

In a *Hustler* article later in the year, Associate Professor of History Paul Kramer called a class he'd taken with the center over the summer "fantastic."

"They really just brought us up to speed and were really patient with faculty, some of whom can be technophobic," Kramer said.[111]

The training paid off. In the end, undergraduate students returning in the fall were offered almost an even split of in-person classes and online courses.

On August 12, as faculty members reached the home stretch of their planning period, they received an email from Chancellor Diermeier and Provost Wente. The message conveyed three goals: "to protect the health and safety of the Vanderbilt community as much as possible; to preserve the core of our academic mission; and to stimulate long-term progress so that we can emerge from the pandemic stronger and more united than ever before."[112]

The message noted recent successes too. Students in the MD program at the School of Medicine had returned to in-person classes in July. And there had been steady and heartening progress in the phased research ramp-up. With a familiar combination of caution and optimism, the message urged faculty "to step up and do our part. Now is the time to set an example for our community."

Supporting Staff Working Remotely

The return of students would not mean a return to campus for most of Vanderbilt's more than 4,400 staff members. Throughout the summer, the university continued to develop ways to support a workforce that had been "commuting by Zoom" for months, often sharing their workspaces with family.

In July, Vanderbilt Human Resources launched three workshops providing training and resources for remote workers. When Metro Nashville Public Schools announced that its academic year would begin remotely, the university created a working group on schools and childcare that was charged with assessing Vanderbilt parents' needs. Made up of faculty, staff, students, and postdocs, the group examined childcare structures already in place at Vanderbilt while also exploring ways to expand them. Working parents were surveyed to help better determine their needs, and the university created a message board where parents could exchange information and support regarding childcare.

Winning Hearts and Minds

Steve Ertel and his team were working fast to ready the "Anchor Down. Step Up." campaign for its scheduled July launch. They'd already designed signs and booklets outlining safety protocols and produced a Return to Campus video, and they'd worked with Nashville's Anderson Design Group to create illustrations for posters, stickers, T-shirts, and other campaign swag, testing early designs with groups like Vanderbilt Student Government and the Graduate Student Council. The team had taken Dr. Hickson's advice about humor to heart: Some materials featured photographs depicting the statues that dotted campus sporting disposable masks. Even the bronze likeness of the Commodore himself, Cornelius Vanderbilt, was now masked on the home page of the university's website.

The campaign launched in July with a social media blitz, a website of its own, and thousands of Vanderbilt-branded face masks distributed to faculty, staff, and incoming students. At the center of the kickoff was the anthem video, featuring Chancellor Diermeier, Athletic Director Candice Lee, Vanderbilt

University police sergeant Shaneithia Lewis, and Ingram Commons dean Melissa Gresalfi, among other recognizable campus personalities, exhorting students to "anchor down and step up" against a soundtrack featuring power chords and the kind of chant-and-stomp often heard at football games.

For the remainder of the summer, Ertel's team finished producing the posters, banners, and signs that would keep up the steady drumbeat of campaign messaging and protocol reminders all over campus—no small endeavor, as it turned out.

"I don't think people understand how manual a lot of these projects are," says Daniel Dubois, who was quick to praise the Plant Operations team for its help. "There isn't a hotline for these things. It's easy to think 'Okay, let's put some light pole banners up all along 25th Avenue.' But someone has to actually go out there, count them, measure them, and get it done."

Where many of the signs for public spaces featured bold, spare design and to-the-point copy, "we understood that more clinical-looking designs would not make sense in students' homes," Dubois says. "So we did what we could to soften the designs for those spaces." The solution? Squirrels. Residence halls' signs featured cartoon squirrels based on the familiar and mischievous rodents that dart up trees and across lawns all over campus and that are the university's unofficial second mascot. "Animals are a great proxy for conveying difficult themes" like mental health, Dubois adds.

On one sign, a masked squirrel pointed disapprovingly at an unmasked pal, beneath the message: "Don't be that squirrel. On or off campus, mask up." By employing "squirrel shaming" as a stand-in for students calling each other out over not following the rules, Ertel's team was able to convey a message about accountability with a lighter touch.

The Return of Commodore Football

Along with the rest of Vanderbilt's departments, Athletic Director Candice Lee and her staff were aiming for a resumption of activities in the fall, but the sports programs were working with a forecast that was considerably murkier. Vanderbilt had been the last university in the SEC to announce it would invite students for in-person classes in the fall.[113] Seasons for various sports were moving targets—particularly football. Still, in early June, Vanderbilt's football players were the first student-athletes allowed to return to campus. With the most players and coaches of any sport, the team carried the highest risk of spreading the virus, so players were brought back in phased groups. "It was all very, very meticulous," Lee says, noting that players were held to the same testing requirements—including tests before and after arriving on campus—being discussed for undergraduate students returning in the fall.

The football players living in or close to Nashville made up the first group to return. They wore masks during workouts, maintained a strict physical distance indoors and out with the help of strategically placed cones, and watched workers spray down the entire football facility with disinfectant every day or two. Conditioning and agility drills normally held inside were moved outside. Weight lifting still happened inside, but only in small groups, which created its own set of challenges.

"One of the things we learned was that it made for some very long days for our strength and conditioning staff," Lee says. "We got to a point in keeping the groups small that they were here for, like, thirteen hours a day starting at five a.m."

The plan was for the football team to go through two to three weeks of voluntary weight lifting and conditioning, with a staff limited primarily to strength coaches, equipment managers, and athletic trainers. July would feature a ramp up to "position-specific" activities, which would also allow more coaches to be present. Preseason practices were scheduled to begin in August.

Tight end Ben Bresnahan and his teammates heard plenty of speculation that the season wouldn't be played, but they wanted to do everything they could to prepare anyway.

"I remember guys like [ESPN's Kirk Herbstreit] saying, 'I don't know if we'll play, it's not looking good,' and all that kind of stuff," Bresnahan says. "There was a lot of uncertainty, but we just went back to controlling what we could control. We were going to do everything we could to prepare to play a full season. If it didn't happen, it didn't happen. But if it did, we would be ready."

In late July, the SEC made two significant COVID-related changes to the league's football schedule—moving the start of the season out three weeks to September 26 and eliminating nonconference games. SEC commissioner Greg Sankey explained that delaying the season would help ensure a safe and orderly return to campuses across the conference.

Then, in August, just as football practices were about to begin, Commodores football head coach Derek Mason revealed that several players had opted out of the season due to coronavirus concerns. Kicker Oren Milstein had been the first to announce his decision, and others, including star linebacker Dimitri Moore and offensive linemen Cole Clemens, Jonathan Stewart, and Bryce Bailey, would follow.

In the weeks leading up to the return of football practice, the school had given its student-athletes as much information regarding COVID-19 as possible. That included a full rundown of the virus's potential effects—like myocarditis, an inflammation of the heart that can weaken it and lead to more significant health issues, and which was a much-discussed concern that summer among collegiate conferences nationwide. All student athletes who made the decision to opt out of the season were able to keep their scholarships, no questions asked.

"Our kids are looking, listening, and paying attention and making decisions for themselves, and that's exactly what me and my program encourage," Mason told ESPN. "You want these guys to be part of your program and part of your team, but not if they're uncomfortable." Moore eventually returned before the season began, but about ten Commodores chose to opt out by the end of the year—more than on any other SEC team.

Bresnahan says he and his teammates appreciated the school's thoroughness in delivering the newest medical information.

"Everyone was very transparent," Bresnahan says. "'This is what it is, this is what we know about it and we're going to keep you up to date on it if we know anything else. It's your decision, one hundred percent.' And they mentioned a bunch of times—about scholarships—that we're not going to look at you differently if you choose to opt out."

Operating an athletics program during a pandemic came back to the school's purpose and mission, says Daniel Diermeier: "Universities are in the business of realizing human potential. That means we want our faculty and students to do what they want to do, and that includes our student-athletes. If they wanted to compete, we would bend over backward to make that possible. We didn't make any judgment, saying 'This is worthy, but that's not.' That's not what we did. Our judgment was, if you committed to something and if this is your passion, your dream, we're committed to helping you get there, whether it's playing football or practicing the clarinet."

All through the summer of 2020, Vanderbilt readied for the task that many said could not be accomplished. Dawn Turton describes the process as methodical, but also "completely exhausting." She remembers visiting classrooms at the Law School being used to pilot the safeguards that would eventually be implemented in hundreds of classrooms across campus—and realizing how long that process was going to take. "And it just seemed like, 'How can we ever be ready?'" she says. "There were certainly times when I second-, third-, and fourth-guessed our decision. And right until the time we opened, I was also fully prepared for us to pull the plug and say, 'No, it's not going to work.'"

By the time August arrived, Vanderbilt would be one of just a small number of the nation's elite universities to invite all undergraduates to campus. To be sure, the fall was going to be a different experience for everyone. Phase II+ of the Return to Campus Plan was now in effect. While it allowed for on-campus residential living, activities, and education, and for some additional increase in research capacity, it limited any gatherings to ten people. Visitors to campus would be mostly prohibited, and buildings normally open to the public, like the library and fine arts gallery, would be closed. Everyone on campus would be masked, distanced, and otherwise following safety protocols and guidelines that were stricter than Metro Nashville's;

consequences for failing to do so included being suspended from the university.

Even with the US coming off a grim summer surge in COVID-19 cases, Vanderbilt's administration was heading into the semester with a measure of confidence, based on the small but encouraging data set it had compiled as the university had slowly "turned up the dial" since May. By mid-August, more than 2,500 faculty, staff, postdoctoral fellows, and students were engaging in on-campus, in-person research. About five hundred School of Nursing and School of Medicine students were already on campus for experiential learning, and the med school had piloted in-person classes starting in July. Whenever officials had spot-checked adherence to safety protocols, they had found excellent compliance. None of the small number of COVID-19 cases that had occurred among people working in the labs had been traced to campus exposure, and response to positive cases on campus had been quick and effective. Throughout it all, communication to the university community and the public beyond had been frequent and thorough. What was more, predictive modeling by university and VUMC faculty showed levels of virus in Nashville and on campus that would be manageable through the fall, further indicating that the campus could be reopened safely.

In another positive sign, admissions had not suffered. More than two hundred additional first-year students were enrolled at Vanderbilt in the fall of 2020 than in fall 2019. The expectation that increased numbers of admitted first-year students across the country would defer enrollment or opt out of college altogether (sometimes late in the summer) led Vanderbilt and other schools to offer admission to more students than usual. But in the end, the incoming class was larger—a strong signal that most students, given the choice, preferred on-campus instruction.

After working practically around the clock since March to safeguard its people and adapt every aspect of its mission-critical operations to create the greatest possible chance of reopening, Vanderbilt's leadership, faculty, and staff were finally ready to test their conviction that continuing with residential learning and research was worth the risk, and that the risk could be managed. In the end, there was only so much they could do. There could be confidence, but there could not be certainty. After all the risks and variables had been examined, reexamined, and mitigated, the only way to find out if a university could fully be a university during COVID-19 was to come together in common purpose and try.

Back in May, in a rush of inspiration, Steve Ertel had, in about ten minutes, written the "manifesto" that was at the center of the "hearts and minds" campaign anthem video. Now, on the verge of a semester rich in both promise and unknowns, it was Vanderbilt's COVID-19 credo:

In the face of extreme challenges, our community is resilient.
Our students and faculty and staff have stepped up.
We adapt. We innovate. We lift each other up. That's who we are.
And we're not done yet.
Our community and our world need us now more than ever. They need our desire for knowledge. Our tenacity to solve problems. They need our compassion and our care.
We are willing to sacrifice for the greater good, to step up and put "we" before "me." That's the Vanderbilt Way.
It's up to each of us to help protect our community.
Anchor Down. Step Up.

PART 3

STEPPING UP

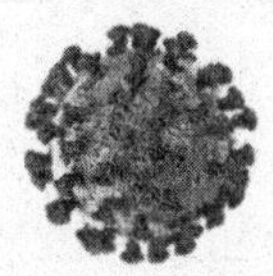

Forward ever be thy watchword,
Conquer and Prevail.
Hail to thee our Alma Mater,
Vanderbilt, All Hail!
—Alma Mater

On August 24, 2020, the first day of the most unusual fall semester in Vanderbilt history, sociology associate professor Shaul Kelner polled the masked and distanced students scattered around his classroom and asked how they felt about being back. Two words came up repeatedly: "excited" and "grateful."

Kelner shared his students' sentiments. But he also wondered how long their good luck would last. "In the beginning, we thought we'd make it for two weeks and then move online," he remembers.

Six thousand undergraduates had returned for in-person classes. Another thousand, including international students with visa complications, students with health concerns, and others, were attending classes remotely. Classes had begun after a move-in process for the more than 3,500[114] undergraduates living on campus that had been considerably more subdued than in typical years. Move-In Day for first-year students, especially, was normally a spirited event with crews of student volunteers in matching T-shirts cheering, throwing the "VU" sign (a peace sign plus a thumb), and helping students and their families hustle suitcases, storage bins, and bicycles into their rooms. Traditionally, even the chancellor was on hand with high-fives and a hearty welcome.

But Move-In 2020 was "silence," recalls biochemistry major Dana Herman, a resident adviser at Stapleton House in

Branscomb Quadrangle. "My floor was a lot quieter, and that was hard initially. It was just a very different level of energy compared to how Vanderbilt normally is." Students were issued their room keys only after passing symptom screening at one of five locations around campus. Arrivals were staggered to avoid congestion, and each student was allowed only one family member or other helper.

"Normally [resident advisers] do a lot, like greeting parents," Herman says. "And this year my role was basically nothing. I made a welcome letter and I dropped it off at people's rooms, but I didn't really go up and interact with them." To help make the transition to college less lonely, each first-year student was assigned a "virtual roommate" to get to know online.[115]

Senior Hunter Long had road-tripped to Nashville with his girlfriend from his hometown of Austin, Texas, excited to return to Vanderbilt for his final year. Long had considered the prospect of more online instruction so unappealing that he had planned to take a gap year if the university hadn't invited students back. Still, he was uneasy about move-in day because he'd heard about "all these rules and regulations." He was anxious to see whether the university could actually pull off what it had been describing in its Return to Campus communications all summer.

"It actually went surprisingly well, which I didn't expect at all," says Long, a photographer for *The Vanderbilt Hustler* who was double-majoring in molecular biology and medicine, health, and society. "I didn't see a lot of people moving in. But it was well organized. It was very uneventful, in a very good way. It wasn't super celebratory, but it was what it was, and it was effective."

One thing Long worried about was helping his girlfriend retrieve her belongings from storage. Her possessions, like many others, had been quickly packed and stored after students had been sent home in the spring. Who knew where they'd ended up or what condition they were in?

"I was like, 'There's no way they did this correctly,'" Long says. "But honestly, it was the most chill and normal experience. Simple and easy." As it turned out, his girlfriend's belongings had been stored in a space high atop the Wond'ry. The pair were able to easily retrieve them.

Adjusting to the New Rules

But Long was right about there being a lot of rules on campus. To maintain lower density, students living in residence halls were only allowed access to their assigned residential building and floor, plus laundry facilities. Guests were prohibited. So were small, informal gatherings of any size—the building blocks of dorm life—and students were told explicitly not to host gatherings in their rooms. Socializing—masked, of course—was restricted to outdoor spaces and to indoor common areas, where furniture had been reduced and rearranged to promote distancing.[116]

Herman says it took some time for students to learn how to hold one another accountable to the rules without creating a culture of accusation and suspicion.

"RAs struggled with feeling like they were mask police. We had to figure out how to have student accountability without turning it into 'students against students.' I think we got a really good balance. People started to see that we were all in this together. As the year went on, people were collectively in support of it more and more," she says.

Even before classes started, Vanderbilt's administration had made expectations clear. In early August, the University of Notre Dame and several other schools offering in-person classes made headlines when students gathering at parties and bars off campus drove increases in infection rates and forced classes online. On August 19, both the University of North Carolina and North Carolina State reverted to remote learning after similar spikes.

"If you ask me when was the moment that I thought this may not work, that was it," Chancellor Daniel Diermeier remembers. "It was make or break."

To get out in front of any mass flouting of rules, Diermeier and Provost Susan R. Wente immediately posted a stern joint message on Instagram that did not mince words.

"Every student must know that the parties and disregard for face masks, physical distancing, and gathering size causing other universities to abandon in-person classes will not be tolerated at Vanderbilt," the message began. "One person's decision to shrug off their responsibility for a night of fun can be the reason an entire class misses its senior year, or why a student, for whom Vanderbilt is the safest home they know, is forced to leave. You must take personal responsibility for the safety of your fellow students on- and off-campus, and for our dedicated faculty and staff who are devoted to your success, and you must prioritize what is important. All students should know that flouting public health requirements in Nashville doesn't just result in student conduct violations—it can carry real criminal penalties.

"We write this not to scare you, but to be perfectly plain: The situation happening at other universities can be avoided at Vanderbilt, but only if you anchor down, step up, and do your part."[117]

"It was Mom and Dad laying down the law," Diermeier says of the message.

Meanwhile, the university's gatherings policy was eliciting objections from a surprising quarter: some students' real moms and dads.

"'Why don't you let them have large social gatherings? Why are you cracking down on parties off-campus?'" Wente remembers parents asking on the Vanderbilt Parents Facebook page and elsewhere. "It just baffles me, as a parent myself. It's almost like they were giving their kids permission to get COVID or something. And we wrote them. We told them we needed them to step up."

On the first day of classes, Diermeier followed up on his and Wente's admonition with a video message that was equally frank but also, in the spirit of a pre-game locker room talk to an underdog team, encouraging.

"Our ability to succeed depends on your choices. Our plan will succeed or fail due to how you will act each and every day," he said in the video. "Make no mistake, there are many who doubt you can do it. You find them on op-ed pages, social media, and occasionally in my inbox. They say, 'You can't rely on college students. College students won't hold each other accountable. Eighteen- to twenty-two-year-olds cannot be trusted to do the right thing.' My call to you today: Prove the doubters wrong. Prove they have underestimated you. Prove that we are One Vanderbilt community that supports one another. Prove that we can step up to any challenge and lead. Prove that, together, we can make this our proudest moment."[118]

Diermeier predicted in his message that "not everything would go perfectly the first time" during the semester, and he was proven correct early that day when students and faculty couldn't log on to Zoom between about 8 a.m. and 1 p.m. and had to improvise with conference calls and alternate video meeting programs.[119]

Busy with Testing (Not the Typical Kind)

The chancellor had also counseled students to expect "moments of normalcy mixed with moments of adaptation and change." One of the biggest changes in life on campus was the mandatory testing program for undergraduates. Near the end of the first week of classes, the university reconsidered its policy of random asymptomatic testing to monitor the prevalence of the virus on campus. On the morning of August 27, Wente sent an email to students and their families announcing that weekly tests would now be mandatory for all undergraduates receiving instruction on campus because universal testing during the week had revealed many asymptomatic positive students.

"This weekly testing will allow us to better and more quickly identify asymptomatic positive cases and localized clusters within our campus community," Wente wrote.[120] One such cluster had occurred just before classes began, when several football players tested positive, prompting the university to pause football activities until the end of the month.[121]

Students quickly fell into the new testing routine, scheduling appointments through the online system they used to register for classes. Once testing was up and running, it was "pretty quick," remembers student Minna Apostolova, noting that the walk to the David Williams II Recreation and Wellness Center, where the program was housed, took longer than the testing process itself. The space, originally constructed to house an indoor football practice field and a track, was large enough to accommodate multiple lanes of physically distanced students snaking through the facility as they waited to get tested. Testing was simple: students spit into a container, placed it on a designated table, and scanned their Vanderbilt ID on the way out to confirm that they had been there. The university had switched to saliva tests from its original plan to use nasal swabs. It was employing tests from Vault Health, a startup that had begun by developing at-home treatments for men with low testosterone before developing the first FDA-approved saliva test for COVID-19.[122]

Apostolova says that hand sanitizer at the testing center was plentiful, and staff members were always at the ready to answer questions. "I give immense credit to the people running the programs," she says. "It was so reassuring to have this built-in, weekly marker and to know that this testing data exists."

Test samples were sent nightly to a lab in New Jersey. It could take as many as three days for students to get their results, either by logging on to Vault's portal or getting an email alert, although on average it didn't take quite that long. Apostolova and her five suitemates in Moore College had a testing day tradition: whenever one of them got results, they would let everyone know in a group chat.

"It's just a way to be like, 'Hey, this is my result.' Thank goodness it has always been negative," she says.

The university's intake of testing results, as part of its monitoring of the virus among all campus populations, was complex. There were more than a half-dozen intake methods, covering tests not only from the Recreation and Wellness Center but also from VUMC, student and occupational health, and the many people who self-declared results of tests taken in the community.

When a positive test result came in, Andrea George's team would prepare a profile of the person who had received it. In addition to contact information and details about where the person lived and worked, the intake team was also on alert for any other telling data: Was the individual a student-athlete, recently exposed during a game? Were they a student at the Blair School, subject to particular risks of choral or wind instrument performance? Did they live closer to downtown Nashville in an environment less shielded than Vanderbilt's campus? Every bit of information they could glean expanded the university's understanding of who was getting the virus and how.

All information was then quickly relayed to the contact tracers, who initiated personal interviews with the infected person and their close contacts, defined by the CDC as individuals who were within six feet of an infected person for at least fifteen minutes within a twenty-four-hour period.[123] Composed almost entirely of nurses and nurse practitioners from the Vanderbilt School of Nursing, the tracing team also provided guidance on testing and quarantining or isolating. Contacts were always notified by phone, with a follow-up in writing as a record for the university—and which would be sent, as applicable, to the dean of students, the provost's office, and school deans.

"There is a huge notification structure, and everything is tracked in a database," Andrea George explains. "Plus, people are often upset and scared when hearing this news—they only hear half of what is said to them. So it's good to have it in writing."

Tracing team members worked twelve-hour days, seven days a week throughout the semester, recording every exchange in REDCap, a tool created at Vanderbilt in the early 2000s to support secure data collection for clinical researchers.[124] In developing the program, Pam Jones and her team had looked at the practices of several universities, some of whom were partnering with their local health departments for tracing. At those schools, the length of time between a student getting a positive test and getting a call from a contact tracer could be as long as five days. Vanderbilt's system was designed so that tracers talked to persons testing positive within twenty-four hours, so as to get close contacts into quarantine more quickly and reduce potential spread.

Training, consistency, and scripting were key to successful contact tracing, says Jones, George's co-commander who led the contact tracing program with VUMC nurse practitioner Alyssa Miller. The team wrote a carefully worded script that tracers would recite in their conversations with contacts. Equally important was the team's warm, human, "high-touch" approach.

"We spent a lot of time with students, faculty, and staff explaining the 'whys.' If people understand the logic behind why they have to do something, it makes a big difference," Jones says.

During the last two weeks of August, 68 of the 4,232 students tested turned up positive. There was an uptick after the first full week students were back on campus, stemming from several small clusters of infections—among lacrosse players, football players, and members of the Sigma Chi fraternity—that accounted for nearly 60 percent of the positive cases. From these outbreaks, administrators inferred that infections were largely originating off campus and in homes with three or more bedrooms. In response, the university implemented shelter-in-place orders for affected groups.[125] Leaders were meeting daily at noon seven days a week to confer about test results and tracing, as well as communications, security, and students' social and emotional welfare.

A Student Experience Unlike Any Other Year

The university faced the biggest test yet of its ban on outdoor gatherings on August 30, a Saturday night, when between fifty and two hundred students gathered over several hours outside The Ingram Commons. Similar gatherings had happened on previous nights, but this was the largest. Most students wore masks and kept their distance early in the evening but abandoned the protocols as the get-together stretched past midnight. In the days that followed, the university responded by posting Vanderbilt police officers in areas where students congregated regularly and reminding undergraduates that penalties for violating the ban on gathering were severe.

"The minimum sanction...will be suspension for a minimum of one semester; a first sanction may be as severe as expulsion, depending on the nature of and circumstances surrounding the violation," the statement read.[126]

Throughout the fall, to encourage compliance with protocols, Vanderbilt police increased its patrols and designated its community safety officers, normally concerned with security and traffic, as public health ambassadors charged with reminding students of protocols in a nonthreatening way. The past summer's protests against police brutality contributed to the decision to uniform ambassadors—and all CSOs—in polo shirts and khakis rather than the police-style uniforms they had worn previously.

"We knew some people are made uncomfortable by any person in a uniform that is associated with law enforcement," Eric Kopstain says of the decision. "We used the COVID crisis as an opportunity to alter the look and feel of the CSOs to make them appear more welcoming and less militarized."

Off campus, the university also employed "knock and talks" at student residences to encourage them to follow guidelines "and, frankly, to signal that we knew where they lived," Kopstain says.

As the semester progressed, Vanderbilt's faculty and students worked together to define what higher education could be like in low-occupancy classrooms where everyone was masked and distanced, instructors were often standing behind plexiglass shields, and there was a good chance at least one remotely participating student's image was projected on a screen.

Shaul Kelner held some of his classes in the tents that had been set up for dining. "I was really trying to give my undergraduates this sense of 'You are physically here at Vanderbilt,'" he says.

Assistant Professor Kevin Galloway viewed the semester as a kind of experiment.

"I told my students, 'I haven't taught a hybrid class before. And you also don't have much experience with them. So some things are going to work well, and others will have to be reformatted or reconsidered.'" The approach was a familiar one for Galloway, an expert in human-centered design and mechanical engineering. "It's about finding out what works and what doesn't and redefining your hypothesis to make it better next time."

Given the inherently hands-on nature of his design classes, Galloway needed a physical space that was as functional as it was safe. His students sat with twelve feet between them, and Galloway created "big orange circles" to guide him; he could walk along their tangents while remaining six feet from everyone. Sometimes he would also stay behind the shield at the front of the room.

For contact tracing purposes, faculty members were encouraged to use assigned seating to help both them and students remember who was where on any given day. They were also encouraged to take attendance.

In September, the Center for Teaching assembled a group of faculty members for a discussion about the challenges of hybrid teaching and the solutions being developed.[127] Instructors were adapting in creative ways. Susan Verberne-Sutton, a senior lecturer in chemistry, matched in-person students with lab partners who were remote; that way, one person was responsible for manipulating the physical equipment, and the other could focus on observing and taking notes.

Professor of Law Terry Maroney used the collaborative annotation platform Perusall to engage students, asking them to use hashtags in their annotations so that she could better identify patterns and trends to serve as springboards for discussion.

Galloway reported that his classroom was a lot quieter than usual—perhaps the byproduct of students sitting so far apart. But Kelner was happy to discover that students could still hear and understand despite the masks and the distance. "They had discussions, they understood each other. It was surprisingly normal," he says.

Not that teaching in the environment altered by COVID-19 was easy. One drawback was the additional procedures instructors had to perform just to start each class. There was setting up the webcam and microphone, making sure the internet signal was strong, opening Zoom, and remembering to make sure in-class and remote students could see PowerPoint presentations and other materials. Galloway found he was constantly checking his laptop screen to see how remote students were doing.

"There was always this back and forth, making sure you didn't forget anybody," he says.

Some faculty also missed informal watercooler conversations with colleagues.

"When I'm in the lab, I can leave my office for five minutes and have a conversation in which we come up with a new idea for an experiment or a different way to look at data," says Alyssa Hasty, professor of molecular physiology and biophysics. "Such spontaneous conversations were less likely to happen on Zoom or over the phone."

Galloway agrees: "Everything is more intentional now." He says that before the pandemic, the Wond'ry had a lot of unexpected drop-ins from visitors with questions and ideas—what Galloway's team calls "creative collisions."

"They may be entrepreneurs or parents or what have you, but they all have their own experiences and passions. And they would share them with us, and they would inform our projects."

Students bumped up against the new restrictions as well. Classroom discussions could be stilted, and being in class could feel like being with a dozen masked strangers pursuing independent learning instead of being part of any kind of community. The learning environment outside the classroom could be quite different too.

"You don't interact with your peers or have office hours in the same way," resident adviser Dana Herman says. "I remember, when I took organic chemistry, going to my professor's office for an hour and there would be like twenty of us huddled in there and trying to learn. And now everything is online, so you're not doing that. Also, places where I would normally study—you're either not allowed to study there anymore or you have to spread out, so there is less room. So you end up studying in new places."

Despite the drawbacks, the semester's initial trepidation gradually gave way to a sense that, overall, classes were working. First-year political science major Chase Mandell discovered that taking advantage of office hours over Zoom "was still a great way to get to connect with [professors] one-on-one and still feel connected to that classroom environment."[128]

And some instructors were discovering unexpected benefits in online teaching or finding that the pandemic lent itself to the subject matter. With the shift to remote learning in March, Tucker Biddlecombe, associate professor of choral studies at the Blair School of Music, had found that students' ability to watch themselves conducting via Zoom helped them improve. Benjamin Legg, a senior lecturer of Portuguese, noted that his intermediate language class was learning about the future subjunctive, a verb construction used to imply uncertainty for the future. "Never before has grammar seemed so eerily appropriate to describing our state of affairs," Legg says. "While our conversation practice could have gotten dark, students in fact made a variety of optimistic sentences." And Jon Meacham, Pulitzer-winning historian and Carolyn T. and Robert M. Rogers Professor of American Presidency, turned Zoom and the curtailed travel schedules of politicians to his students' advantage by inviting US House Speaker Nancy Pelosi and US Secretary of Transportation Pete Buttigieg—even former president George W. Bush—to be virtual guest lecturers in a class he co-taught on US elections. The course turned out to be the largest, and one of the most popular, in Vanderbilt history, with nearly 850 students enrolled.

When several weeks had gone by without the university having to suddenly switch back to all-online learning because of an outbreak of the virus, Kelner says, "It was like, 'Oh, we're actually going to win this thing.'"

One day in August, a photo from a colleague appeared on the phone of Vice Provost for Academic Affairs and Dean of Residential Faculty Vanessa Beasley. It showed three first-year students sitting in the grass near the Ingram Commons Center, masked and distanced, but obviously getting acquainted. Beasley was touched by "that ordinary act, of new students meeting each other and getting to have that first real talk after their families had left. It was something so simple, yet still such a big first step. And it was still possible in a pandemic."

As autumn came on in earnest, Vanderbilt's undergraduates did their best to supplement their class time with something resembling a residential college experience. For some first-year students, like human and organizational development major Macy Su, just being on campus and getting a taste of independence and community was satisfying—and an improvement over life under COVID-19 as they'd known it.

"I came into college having missed out on some important events like prom and graduation. So I think when our class got here, we were so eager to meet each other and make connections because we lived through the remainder of our senior year not having that," she says.[129]

The residential staff and faculty head of Su's house in The Ingram Commons were creative in the programming they offered to help students relax and connect. There were trivia nights and paint nights. Other halls put on virtual hobby nights, where students could log on to learn a new skill. There were online interest groups for gaming, jewelry making, and jigsaw puzzles. There were in-person activities for small groups, including high-intensity interval training, yoga, gardening, and faculty salons. There was a Class of 2024 scavenger hunt. There was "speed-friending," in which students met outside, masked and distanced, to meet new people in ten-minute blocks. There were gatherings at fire pits and outdoor movies.[130]

"This version of Vanderbilt is all we've ever known, and I can't imagine it any different," Su says. "But I'm looking forward to a time when we can go to in-person events and do things like sit together in a group at a football game."

Some older students who'd experienced Vanderbilt pre-COVID were acutely aware of what they were missing.

"Pretty much everything feels different," Herman says. "There's really no social events or parties, none of that extracurricular stuff. It's all online. Dining is really reduced, or it's weird because you can't really sit with your friends; you're either sitting outside in the tents or, like, sitting in a booth alone."

An important part of the job for Herman and other RAs was keeping an eye on residents' mental health. "People spend so much time in their rooms. They don't have many opportunities to meet people, so that's been a really big thing. . . . Getting people to bond is challenging."

With options for socializing limited, some students paid more attention to fitness. Students on Herman's floor mounted pull-up bars in the doorways of their rooms. Some days, the hallway would be lined with students hanging from the bars, straining, sweating, and screaming.

"It was good to see people, and we would play music to make it more lively," Herman says. "It was interesting to see, as the fall progressed, just how creative people were. . . . It was really weird at first, for sure. But then I think it kind of got a little bit better as people got more used to what we could and couldn't do. . . . I think a lot of people honestly don't still know the full extent of how people our age are reacting when you're so used to being social 24/7."

For senior Hunter Long, a baseball fan, a bit of saving grace came in the shape of a ball.

He and his three suitemates passed time by watching ball games, sometimes all day. They took breaks to play catch on the lawn outside their hall. They weren't alone.

"There were a lot more people outside than there would be normally on even a super nice day. The year before, there might be a couple of people outside. But now everyone was out. That's one of the most interesting things from the pandemic that I will always remember: tossing the baseball with my friend. Just doing that, and having the flexibility and freedom to be outside. You definitely take a lot less things for granted. A lot less."

The housing plan for first-year students that was developed over the summer seemed to be working out. Despite some initial feelings of disconnection, the first-year students living in Carmichael Towers began carving out their own community.[131] Some were even inclined to skip "the Flip."

"Once we got it going, we had students in the Tower saying, 'I don't want to flip. I built my community here and am happy to stay,'" says Senior Director of Residential Experience Randy Tarkington.

Feeding students required a bit more problem-solving on the fly—what Sean Carroll in marketing calls "continuous adaptation and reinvention." The week classes started, six dining halls were up and running. Mobile ordering was also available from fifteen locations across campus, including a kosher food truck, a coffee shop, and several of the Munchie Mart convenience stores around campus. More food trucks arrived in September, and the university's Taste of Nashville program, which allowed students to use their meal plan money at restaurants, expanded to more than forty locations. Demand was high for "market baskets," a new initiative offering groceries like grains and fresh fruits in lieu of a prepared meal. On-campus markets had changed to an order-ahead model to reduce the number of students packing into the stores and browsing.

By the end of August, adjustments had been made to address a few kinks in the system. Because demand for space in the dining tents was expected to be high, tents had been secured with fencing and access to them was controlled using a text-based reservation system. The controls worked a little too well—few students dined in the tents initially. But when dining services staff removed the access controls, use increased. Also, many students who'd avoided the tents had taken to eating on the ledges of low masonry walls throughout campus, and they were not always mindful of the required six feet of physical distance. All-weather stickers affixed to the ledges, indicating where students should sit, solved the problem. And in the dining halls, physical distancing was leading to long lines. Shifting briefly to boxed grab-and-go meals, with a limited number of choices, reduced wait times while dining staff revised the queueing system.[132]

"It was really hard to go from dining, which I really loved, to getting handed boxed lunches every time," Long says. "I would say it was demoralizing. It was one of the worst experiences and the biggest shift. To not have any choices was the hardest thing."

With so many adjustments to menus and operations throughout the semester, social media played a key role in disseminating information. Dining staff also texted alerts of changes or new features to students' phones.

A consistent backdrop of the student experience was the Division of Communications' ongoing "Anchor Down. Step Up." campaign. Throughout the fall, new signs featuring the now-familiar cartoon squirrels appeared around campus with encouraging messages like: "You're doing great, Vandy. Keep masking up." A sign featuring a squirrel modeled after Rosie the Riveter, complete with the hair scarf and flexed bicep, read: "We can do it! Stay strong to stay together."

The communications team introduced new elements of the campaign, including more than four hundred posters with the headline "Who do you mask up for?" and featuring photos of students, faculty, and staff members holding whiteboards with their handwritten answers: family members, front-line workers at VUMC, essential workers on campus. Hanging in locations where their messages would resonate most, the posters were a moving reminder of the lives students and others on campus were protecting by following protocols.

College Sports Return

With the return to campus came the return of collegiate sports. One student-athlete grateful for that was senior and women's tennis player Christina Rosca, who had feared her collegiate career had come to a premature end when the SEC suspended athletics in the spring. But Rosca learned while at home in New Jersey over the summer that the NCAA, in light of the pandemic, had granted an extra year of eligibility for 2019–20 winter and spring student-athletes, which allowed her to return to Vanderbilt for another year.

"I was really happy that I would get a chance to finish up. I'd been playing the best tennis of my life in March when we got canceled," she says.

SARS-CoV-2 infections continued to hound some of Vanderbilt's teams. In early September, more football players and a number of soccer players tested positive.[133] (In the end, it would eventually be established that, contrary to concerns before the fall sports season, no infections among athletes were traced to the field of play; transmission appeared to occur during social interactions off the field.)

With the start of football season delayed by SEC directive, Vanderbilt's first competitive athletic event of the fall—and the first since spring sports had been canceled—took place on September 23 at Nashville's sprawling and wooded Percy Warner Park. There, the men's and women's cross-country teams hosted the Commodore Classic, drawing squads from Alabama, Auburn, Kentucky, Missouri, and Tennessee. It was a pared-down event, featuring about forty runners instead of the usual two hundred or so. But it felt significant to Athletic Director Candice Lee, who alerted her colleagues in the athletics department that morning with a simple message: "Y'all, we're back."

"I was so excited," Lee says. "Because you felt like, 'Oh man, this is really going to happen. We're really going to get a chance to compete.'"

At the same time, the football team was preparing for the season opener it would play a few days later. The roster looked different because of the players who had opted out in August. And looming over the Commodores program—and programs nationwide—was concern not only about more infections but also about myocarditis, an inflammation of the heart that was sometimes a complication of COVID-19 and which, in serious cases, could cause permanent damage or a heart attack. Given the strain athletes regularly put on their hearts, myocarditis could be especially dangerous.

"That was a really scary thing, scary information that was constantly evolving," Lee says. "As we learned things, we shared very openly with student-athletes, with parents, with coaches, with staff, and we kept those lines of communication open."

Commodores football began its season with a road contest at Texas A&M University that required a two-and-a-half-hour flight from Nashville. To limit the chances of infections, Lee reduced the numbers of coaches, trainers, and athletic department personnel on the trip, trimming dozens off the 125 or so people who usually traveled to games. No guests, family members, sponsors, or donors were allowed. And to prevent the virus from spreading through food and drink, there was no catering on the plane. Instead, individual bags in every seat contained bottles of water or Gatorade and straws. Later in the season, when the Commodores hit the road for games closer to Nashville, they used nine buses instead of the usual five or six to allow for more spacing.

The football team played its first game of the season on September 26 before a crowd of about 24,000 at College Station, taking a 17-12 loss to the tenth-ranked Aggies. There would be nowhere near as many fans in the stands at Vanderbilt Stadium the following week, when the Commodores hosted Louisiana State University's Tigers in the team's home opener. Each school in the SEC had been granted the freedom to set its own attendance policies. While the other thirteen league members had set their caps between 20 and 25 percent of full capacity, Vanderbilt chose what administrators felt was a safer path. Lee had announced on September 11 that games would be played without any spectators through October. She amended the policy a bit later, allowing limited student attendance starting with the game against the Tigers. Knowing the decision to limit attendance might be met with questions by her student-athletes, Lee wanted to deliver the news personally. She walked onto the soccer and football fields during practices one afternoon and explained her reasoning. She told players she knew how badly they wanted to compete. And as someone who felt responsible for their well-being, she also wanted to choose the safest option when dealing with a virus that was still such a mystery.

"It was a challenging thing to say because I was a student-athlete, and I can tell you, I can't imagine playing without fans," Lee says, looking back on those conversations. "And I felt awful, too, because here I was, a first-time athletic director, making the fan base mad. I mean, you need your fan base. They're the lifeblood of the program. It's important. But I had to prioritize these student-athletes and their health and safety over everything else. It's been hard."

Managing the Day-to-Day

Throughout the month of September, Vanderbilt's positivity rate hovered between 0.3 percent and 0.6 percent, with between twenty and thirty-eight students testing positive week to week.[134] Davidson County's positivity rate during the month ranged between 9 percent at the start of the month and 3.6 percent toward the end.[135] Near the close of September, five students who'd just begun isolating after receiving positive test results were notified that their tests were, in fact, negative. Several of their close contacts—including three who were resident advisers—had been placed in quarantine. The mix-up was the result of a communications error by test provider Vault, which promised fixes to its reporting system.[136]

That there was any normalcy on campus was something of a miracle, given that, as September turned to October, thousands of the staff who made the wheels of Vanderbilt turn daily were still working from their dens, spare bedrooms, kitchen tables, and couches in homes scattered around Metro Nashville. Everything that happened in Vanderbilt's classrooms, dorms, and athletic spaces was made possible in part by a parallel university operating over the internet. Working from home month after month could be far from ideal, especially for staff sharing space with family. But like workers all over the world, Vanderbilt's staff were discovering that a workday—and their work-life balance—could look different than it did before the pandemic.

"I'd like to be back on campus because I miss connecting with the students that I mentor and seeing my colleagues, but I do see the benefits, like spending a little more time with my kids and not having to commute and all of those things," says Anna Thomas, assistant director of training and communications for the Office of the Dean of Students.

As a show of gratitude for staff, the university held its first Staff Assembly, modeled after the regular Faculty Assembly traditionally held each spring and fall. More than 1,250 staff members watched online as Chancellor Daniel Diermeier gave remarks and presented Anchor Down, Step Up Excellence Awards to nine individuals and ten teams for above-and-beyond effort during 2020.

"Your hard work and dedication throughout this unprecedented year has exemplified our One Vanderbilt spirit of collaboration," Diermeier said. "The time and effort you put in for our students, faculty members, and your fellow staff is critical, and will continue to be critical, to Vanderbilt's mission."

"We pushed our staff beyond reasonable expectations," Vice Provost for Faculty Affairs Tracey George says, looking back. "Fortunately, but not by accident, we have outstanding and loyal staff who worked hard, stayed positive, and gave it their all despite having to manage the pandemic's impact on their personal lives. Our next task is keeping them and identifying how we can build capacity to shift when the next crisis comes."

The semester was proceeding successfully enough that on October 7, Vanderbilt announced its plan to continue in-person learning in the spring, "governed by our current safety protocols."[137] Then, during the week of October 19, the number of positive tests jumped to one hundred, up from forty-four the previous week, for a positivity rate of 1.54 percent—the highest number of positive results of the semester to that point. Another eighty-two were recorded the following week. The numbers mirrored the autumn surge that was beginning in Davidson County at the time, though they were considerably lower than the county's positive rates of between 6 and 7 percent during the same period.[138]

With each positive test came days of isolation for the person testing positive and quarantine for any of their close contacts identified by the university. Students heading to isolation or quarantine had to quickly pack a few things to take to the designated quarters. "We would always see the carts carrying the kid with the pillow, and it was like, 'There they are, they're going,'" Long recalls.

Some students found the experience to be lonely, marked by empty white walls and microwave meals. Others discovered that quarantine and isolation housing was more comfortable than their normal accommodations—they had full kitchens. Some students were placed with roommates, offering a welcome break from the semester's residential solitude. The university's contact tracers would check in regularly, applying their nursing backgrounds to assess not only students' symptoms but also their mental health while being separated from the rest of campus.

As October came to a close, Vanderbilt's leadership rolled out strategies to prevent large student gatherings over Halloween weekend, when some of the biggest student parties of the year typically took place. Wente sent an email to students encouraging them to "stay strong" and adhere to protocols.[139]

In an effort to provide safe alternatives to downtown Nashville and off-campus parties, the Division of Communications partnered with other campus offices to organize activities like outdoor pumpkin decorating, a movie night—where students sat in hundreds of socially distanced hula hoops in the university field house to watch films that included *The Rocky Horror Picture Show*—and a self-guided audio tour of the Vaughn Home—an 1875 structure that was one of the first houses built on campus and was rumored to be haunted. The Vaughn Home tour included masked communications staff, normally responsible for top-level university responsibilities, gamely costuming up to surprise and frighten student guests late into the evening, in true haunted house fashion.[140] Multiple food trucks were deployed to keep students engaged on campus. Vanderbilt police used their relationships with landlords to discourage parties off campus. The university's efforts worked—most students rose to the occasion and stuck to the rules, and, for the most part, the blowout bashes of years past did not materialize.

In all, 257 undergraduates tested positive between October 5 and November 1. The positivity rate would continue to increase through the end of the semester, reflecting the climbing rates in the county, state, and nation.[141]

Encouraging News on the Medical Side

As worrisome as that was, there was hopeful news coming from the research labs of the university and the medical center. Following the ramp-up to Phase II+ on August 14, pre-approved research activities had been operating at 50 percent capacity throughout the semester.

In October, the biopharmaceutical company AstraZeneca announced that it had begun Phase 3 trials of a COVID-19 therapy consisting of two of the monoclonal antibodies developed by James Crowe and Robert Carnahan and their colleagues at the Vanderbilt Vaccine Center. AstraZeneca's aim was to enable a single dose of the therapy to remain effective in the body for six to twelve months—a potentially powerful benefit for the elderly, for people with compromised immune systems, and for people unable to receive a COVID-19 vaccine.

Later in the month, the FDA approved use of the drug remdesivir as a therapy for COVID-19, building on research by Mark Denison, Andrea Pruijssers, and their colleagues in the Denison Lab that showed the drug to be effective.

"We started this work years ago, knowing that with no therapeutics on the market to treat coronavirus infections, we would be in trouble if a pandemic hit," says Andrea Pruijssers, lead antiviral scientist in Denison's lab. "We are now developing additional drugs to battle COVID-19 and to be better prepared for future pandemics."[142]

And in the biggest news of all, on November 16, Moderna announced that early Phase 3 trials of its mRNA-1273 vaccine, which Denison and his colleagues had been instrumental in developing, had shown it to be 94.5 percent effective, making it one of two vaccines likely to receive emergency FDA approval before the end of the year. Fans of Dolly Parton reacted "rapturously" to the news, *The New York Times* reported.[143]

The encouraging news on vaccines and therapeutics came at just the right time too. Because on November 8, the number of SARS-CoV-2 infections in the United States passed the ten million mark, with a million of those cases added in just ten days. The nation was gripped by the "third wave" of infections that experts had predicted.[144] Between November 2 and November 8, the number of positive COVID-19 tests among Vanderbilt undergraduates reached 110, working out to a positivity rate of 1.66 percent. Those record numbers were topped the following week, with positive tests totaling 117. Still, the positivity rate of 1.7 percent was substantially lower than the 8.5 percent recorded by surrounding Davidson County the same week.[145]

Rays of Light Amid the Darkness

The week of November 9 through November 15 marked the end of mandatory testing for the fall. With students on campus for just five more days of classes, they were given the option of registering for testing between November 16 and November 18.

The testing and tracing teams continued to work almost nonstop through the end of the semester. So did the university's communications team, which, in addition to providing updates on test results and the evolving plans for the spring semester, found that the pandemic was unexpectedly expanding its capabilities for engaging with the Vanderbilt community worldwide. The shift began with the online town halls held earlier in the year. Convening such events using Zoom and YouTube was new territory for Steve Ertel's team.

"We bought a webinar license and figured it out," he remembers.

But as the semester unfolded and the team's grasp of the technology grew, Ertel's team began to understand the opportunities that reaching a vastly larger audience afforded. That realization hit home with the November livestreaming of an installment in the Chancellor's Lecture Series. In-person lecture series events, which featured high-profile guests from politics, media, and the arts, almost always sold out the 1,100-capacity Langford Auditorium—the largest venue on campus aside from the football stadium or Memorial Gymnasium. But the number of people who could watch the November 16 lecture featuring former US secretaries of state Madeleine Albright and Colin Powell was limited only by the parameters of the Zoom license; more than 2,800 people "from all over" attended the event virtually, Ertel says. He and his team suddenly saw exciting possibilities for making events that had once been restricted to campus audiences accessible for alumni, families, and members of the public anywhere there was an internet connection.

The university also pressed forward with its work on diversity, inclusion, and equity, advancing initiatives it had outlined in the detailed July 1 message from Diermeier, Wente, and

Chief Diversity Officer André Churchwell. That message—which Churchwell calls a "galvanizing document"—became the framework for the university's efforts. Subcommittees of the University Diversity Council, which included faculty, students, alumni, and staff, were charged with making recommendations for accomplishing each initiative. The committees were scheduled to share their recommendations in the spring of 2021. Churchwell's office also polled the university's deans to learn what they were doing to promote equity, diversity, and inclusion. This yielded a report of some five hundred pages, from which best practices were culled and shared university-wide. Churchwell says the events of 2020 gave new momentum to a number of actions that had been underway before George Floyd's death and the protests that followed.

"We'd been working on this stuff for years. But at this moment in America? Put your foot on the accelerator. We've got to do more," he says. Churchwell's father had been a reporter with the *Nashville Banner* in the early 1950s—the first Black journalist to work full-time at a major newspaper in the South. He often told his son that the fight for racial equality was Sisyphean, with progress often being lost after a bit of progress was made. After the summer of 2020, Churchwell says, "We've got more shoulders pushing against that boulder."

As the semester neared its end, the coronavirus unexpectedly brought much of the Vanderbilt community together through the heartbreak and triumph of sports and the experiences of two women whose athletic trajectories were altered by COVID-19 in utterly different ways.

On November 17, the women's basketball team canceled its scheduled home opener against Tennessee Tech. The quarantining of players, as well as injuries, had left the squad with fewer than the seven players the SEC required.[146] One of the players who tested positive, sophomore guard Demi Washington, was eventually found to have myocarditis. The tests the SEC required to screen for the condition had all come back negative.[147] But an MRI, which only Vanderbilt required, showed worrisome damage to Washington's heart; she was out for the season.[148] After more delayed and canceled games caused by positive tests, opt-outs, and injuries, the women's basketball team discontinued its season.

Around the same time Washington tested positive for the virus, Vanderbilt's football team was heading into the last full month of its season. Athletic Director Candice Lee announced that the final two home football games would be open to parents and families of players, as well as to a select number of graduate and professional students. The hope was to fill seats in the absence of undergraduates, who would be heading home after in-person classes ended on November 20.

The team had a tough season, with COVID-19 taking its toll through the infections, quarantines, and opt-outs that had depleted the roster and forced postponement of a contest against Missouri scheduled for October 17. But when that game was finally played, it would make history.

In the week leading up to the November 28 game, the team's kicking specialists had been affected by COVID-19, so they were not available to play. Searching for someone with a strong leg, football head coach Derek Mason made a groundbreaking and logical move: he asked Commodores women's soccer goalkeeper Sarah Fuller to step in. Fuller, whose team had just captured Vanderbilt's first SEC women's soccer championship in twenty-six years, had a powerful leg, which she demonstrated during games, clearing the ball with booming goal kicks. Fuller agreed, and Mason and Assistant Athletic Director Jason Grooms relayed the news to Lee in an exchange that was surprisingly low-key, given how much attention Fuller's participation would eventually garner.

"They just sort of passed by and they were like, 'Hey, just a quick heads-up: Sarah—you know, Sarah from the soccer team?—she's going to join us this week,'" says Lee, who knew Fuller had practiced with the team. "And I said, 'Okay, cool.' And it wasn't like, 'Oooh, Sarah!' It was a transactional conversation. I just told them to make sure to check with [the school's compliance department] because it was a student-athlete from one team joining another. That was it."

With Lee's okay, Fuller was on her way to becoming the first woman to play football in an NCAA Power Five conference.

Walking into the hotel lobby in Columbia, Missouri, on game day morning, Lee began to understand that the rest of the world regarded Fuller's spot on the roster as a big deal. A wide-screen TV showed ESPN interviewing Vanderbilt women's soccer associate head coach Ken Masuhr at his Nashville home.

"Then when we got to the stadium and I saw her get off the bus, it really hit me," Lee says. "When I saw her run out on the field with her helmet on, I have to admit it almost took my breath away for a second."

Fuller's big moment came at the start of the second half, when she booted a squib kickoff that was downed without return at the Missouri 35-yard line. Immediately afterward, an SEC-designed video paid tribute to Fuller, and the Mizzou home crowd gave the Commodores kicker a standing ovation.

"It's just so exciting," Fuller told ESPN afterward. "The fact that I can represent the little girls out there who want to do this or thought about playing football—or any sport, really. And I [hope] it encourages them to be able to step out and do something big like this."

Fuller took another step two weeks later in Vanderbilt's clash with archrival University of Tennessee when she successfully kicked a pair of extra-point attempts to become the first woman to score in a Power Five conference game. Days later, the College Football Hall of Fame in Atlanta put Fuller's uniform on display, and she would eventually make a number of national appearances—including one at President Joe Biden's inauguration in January, where she introduced Vice President Kamala Harris to a national TV audience.

Looking back on the semester, Lee says that despite the obstacles, she saw Vanderbilt's student-athletes find their voices and use their platform more than ever. Some of that empowerment came out of the pandemic as student-athletes made the difficult decisions to opt out of their respective seasons. But some of it also came from issues beyond COVID-19, such as the fight for racial equality.

"I think we've seen a shift in student-athletes really authoring their own experience," Lee says. "We're enduring two pandemics: one is about COVID-19, one is about social justice. The social justice piece is, like, athlete activism and fighting against racial inequality. [They're asking,] 'How do I use my own voice? How do I make sure I'm getting what I need? How do I make sure I'm not being treated as labor?' There's a lot to unpack there, so these things are colliding. I would have to think we're going to see the effects for a long time."

Mission: Impossible . . . Accomplished

In-person classes for the fall semester had ended at Vanderbilt on November 20. Most students left campus for home, where they completed a last week of classes and their final exams remotely. The term officially concluded for undergraduate and graduate students on December 13.

It had been a semester like no other, during a year like no other. Vanderbilt's administrators, faculty, staff, and students had done what they set out to do. When much of the world was saying it wasn't possible to safely and effectively teach students on campus or conduct in-person lab research, Vanderbilt had dared to look more closely at that assumption and found it to be false. It had cost the university some $40 million to safeguard the campus. It had required faculty and staff to put in a staggering number of hours and to stretch nearly to their breaking points. And it had required students to choose maturity, prudence, and seriousness in ways that many of their parents and grandparents, if they attended college in a less-fraught time, never had to. By any measure, the semester had not been easy. But Vanderbilt had proved the doubters wrong.

For Daniel Diermeier, the fall semester validated decisions made in the spring, when so much was unknown and the university could only look to its mission as its true north. "We not only managed the challenges of the pandemic in a climate of fear and immense uncertainty, we did it in total alignment with our mission and values. Our entire community—staff, students, parents, alums, and faculty—aligned on a common mission and together overcame the biggest challenge we've had in the history of the university," he says.

For Susan R. Wente, the fall's success was a validation of her "three Ts"—trust, transparency, and teamwork.

"It's all about how we make the decisions, not necessarily what decisions get made," she says. "If I were to pat us on the back, I think we have done a good job sharing that 'why' behind our decisions. And we've done a good job trying, of taking the approach that, if we haven't gotten it right the first time, we'll try again."

Dawn Turton says 2020 provided a case study of how Vanderbilt's collaborative culture enables it to take on big challenges.

"Higher education talks a lot about being collaborative. But I've been in higher ed now for a long time, and I have never seen collaboration to this degree. We've deployed people who would have never worked closely with each other, and everyone just said, 'Let's roll up our sleeves and get it done.' They put egos aside. The workdays became eighteen-hour days, and people just got on with it.

"We didn't do the easy thing," she says. "We didn't do the popular thing. But we did the right thing."

Says André Churchwell, "Our secret sauce—the Vanderbilt collegial environment of working together across the institution—really enhanced our ability to communicate what we're doing, share ideas, and come up with solutions on the fly. As painful as it was, and as challenging as it was, it actually allowed Vanderbilt to shine."

A major victory of the semester was the fact that none of the infections that turned up on campus were traced to classrooms or labs—a great relief to many faculty.

"Our success in the fall has won over a faculty who is supremely intelligent and has extremely high standards in all respects," says Vice Provost for Faculty Affairs Tracey George. "This is the basic feature of institutions like ours—you can try to persuade with words, but the faculty are simply too smart to just accept arguments that they can immediately critique and counter. You ultimately have to win them over by proving that you were right—which puts a lot of pressure on us to be right."

On the last day of fall classes, when Associate Professor Shaul Kelner's sociology students were back at home and attending his class remotely, he grew concerned when, one by one, the rectangles on Zoom where his students' faces had been went black. Before he could troubleshoot the connection, the students began popping back up, each of them holding a handmade "thank you" sign.

"I have never experienced anything like that. It was incredibly special, and I'd say it's one of the highlights of my career as an educator," Kelner says.[149]

After the semester, the editorial board of *The Vanderbilt Hustler* expressed appreciation as well.

"It's important to offer a 'thank you' to Vanderbilt for taking a risk for its students when other universities were unwilling," the board wrote. "It's not exactly in our cynical DNA to express gratitude, but in spite of the constantly evolving, in-need-of-improvement nature of the previous semester, it's a triumph that we were able to receive one at all."

"I'm excited to see how our experience this first year is going to shape the next four," says first-year student Macy Su. "When we graduate, we're going to look back on this time and think, 'We overcame that? We can do anything.'"

For Hunter Long, who was on track to graduate from Vanderbilt in May 2021, being on campus was literally life-changing. He picked up his second major in health, medicine, and society, he says, because of what he saw and experienced during the pandemic—particularly the health and insurance systems he watched "break down." He says Vanderbilt did the right thing in bringing students back.

"Do I think they did everything perfectly? No. Do I think they did everything as well as they could have for being one of the few universities to actually reopen? Yeah. And that's what it boils down to. There are so few schools that did what we did. So to be a small private school that actually reopened—it's impressive. There have been opinions, and people have talked about whether it was actually worth it or not. At the end of the day, for me, it was. As long as there was in-person residential life happening, I was going to be back."

EPILOGUE

The end of the fall semester, of course, marked only the halfway point in the 2020–21 academic year. After Vanderbilt's prolonged winter break, students and faculty would return to campus and to a protocol-bound routine that would seem more familiar and less like a calculated risk. Lessons learned during the fall would be applied to everything from instruction to dining to communications. More students would opt to return, and the university would offer undergraduates more classes taught primarily in person.[150]

But first came December.

With the conclusion of the semester at mid-month, as COVID-19 cases in Metro Nashville and around the nation climbed steadily in a post-Thanksgiving surge, Chancellor Daniel Diermeier hosted an online event thanking the Vanderbilt community for its dedication and resilience. "We're in This Together: A Vanderbilt Show of Gratitude" featured acclaimed Nashville-based singer-songwriter Jason Isbell and his band, The 400 Unit; Vanderbilt singing group Melanated A Cappella; and singer, Blair School alumnus, and national touring *Phantom of the Opera*-lead-turned-COVID-19-song-parodist Chris Mann.[151]

The same day as the show, Vanderbilt University Medical Center began administering its first COVID-19 vaccinations to hospital staff.[152] Despite rising COVID numbers and talk of "a dark winter," the historic acceleration of vaccine development and the vaccines' initial rollout provided a weary world with a light of hope going into the holiday season.

And then, in the early morning of Christmas Day, a disturbed man blew himself up in a bomb-rigged RV parked on a downtown Nashville street, bringing glass, bricks, and devastation raining down on a historic city block and—by damaging a nearby AT&T network hub—knocking out phone, internet, and cable service for thousands of people in several states. Thanks largely to the rapid evacuation of residents by Nashville police, no one other than the bomber was killed. But for many in the Vanderbilt community, the blast was a violent punctuation mark at the end of what was universally acknowledged as an awful year. Members of campus security, information technology, plant operations, and communications teams spent Christmas taking measures in response to the bombing.

January brought continued rollout of two fast-tracked COVID-19 vaccines and, for the Vanderbilt community, thoughts of the spring semester. The university announced that mandatory testing would be increased to twice weekly and that the new testing vendor it had selected would be able to deliver results more quickly. In a virtual town hall for students and their families, university officials described Vanderbilt undergraduates as "the best spitters in the country."

The first month of the new year also brought more unrest, when a violent mob of right-wing extremists, conspiracy theorists, white supremacists, and others loyal to President Donald Trump stormed the halls of the US Capitol to stop Congress from certifying the electoral victory of Trump's opponent, Joe Biden. The day left five people dead and shocked the world. The same month, Professor Jon Meacham, College of Arts and Science dean John Geer, and others launched the Vanderbilt Project on Unity and American Democracy, which aims to bridge America's political divide by "elevating the role of research and evidence-based reasoning in the national conversation" and using examples of unity from American history to inform and inspire greater national unity today.

By February, the "Anchor Down. Step Up." campaign had garnered more than 283,000 engagements on social media, more than 300,000 video views, and nearly 400 orders of swag.[153] The anthem video—which encouraged viewers to put the "we" before "me" in slowing the spread of COVID-19—was a finalist in the 2020 CASE Platinum awards for best practices in communications and marketing.

There was good news from the university's finance office as well. Vanderbilt's endowment had regained all the ground it lost early in 2020 and had continued to grow, totaling more than $8 billion by the end of the year, and growing even more in early 2021. Alumni also continued to step up despite the hardships of the pandemic; the significant drop in giving that administrators anticipated did not materialize.

Also in February, an MRI showed that basketball player Demi Washington's heart had healed from the effects of myocarditis. Whether she'll play again remains an open question.[154] By mid-February 2021, a total of 138 Vanderbilt student-athletes had tested positive for COVID-19 since August 2020. Six developed myocarditis.[155]

In March—nearly one year after Vanderbilt's campus had been largely closed down and commencement ceremonies postponed—Vanderbilt announced it would hold an in-person commencement for the Class of 2021 in mid-May. Students would be allowed two guests each, and guests could sit together in distanced, "pod-style" seating. Speaking at the traditional Graduates Day event the day before commencement would be Dr. Anthony Fauci, the longtime director of the US National Institute of Allergy and Infectious Diseases, who was seen as both hero and villain during 2020 as Americans' reactions to government-imposed COVID restrictions split along political lines. Fauci—who would attend the ceremony virtually—was chosen to receive Vanderbilt's Nichols-Chancellor's Medal, which is awarded to those persons who, in the estimation of university leaders, "define the 21st century and exemplify the best qualities of the human spirit." Not forgotten, the Class of 2020 would finally have its own commencement ceremony. It was scheduled for early May with an address by author, attorney, and former ambassador Caroline Kennedy, the daughter of US president John F. Kennedy.

On March 2, VUMC vaccine research benefactor Dolly Parton received "a dose of her own medicine" when she was filmed getting vaccinated while sitting in front of a Vanderbilt logo wall with her friend, surgery professor Dr. Naji Abumrad. To encourage others to get the shot, Parton revised the lyrics to her classic song "Jolene," singing,

Vaccine, vaccine, vaccine, vaccine,
I'm begging of you, please don't hesitate.
Vaccine, vaccine, vaccine, vaccine,
'Cause once you're dead then that's a bit too late.

Throughout the month, vaccination efforts nationwide steadily picked up speed.

As this book goes to press, Vanderbilt is in Phase II+ of its Return to Campus Plan. Thousands of staff are still working from home, and students around the world continue to take classes remotely. On campus, students and faculty still follow strict safety protocols, and, commencement plans notwithstanding, the ban on gatherings remains. Testing has resumed in expanded fashion, now being conducted twice a week for undergraduate students, with a turnaround of about twenty-four hours. Positivity rates have also dropped and are hovering around 0.3 percent. Meanwhile, thousands of the student belongings abandoned during the sudden March 2020 move-out remain unclaimed. Efforts to return them to their owners continue.

To date, more than 3 million people have died of COVID-19 around the world, according to Johns Hopkins University data; more than half a million of those deaths occurred in the United States. Average life expectancy in the nation dropped by a year during the first half of 2020, and, according to the CDC, Black, Hispanic, and Indigenous people in America continue to die of the virus at disproportionate rates.

There will never be a doubt that this has been one of the most devastating periods in recent memory for all of humanity.

But through these hardships and heartaches, Vanderbilt has endured as a community of strength, compassion, intellectual talent, and resilience—driven by an unshakable belief in a bright future ahead for the university and for the world.

LESSONS LEARNED

WEATHERING THE STORM

Chancellor Daniel Diermeier

As the book's title highlights, this truly has been a year like no other. The world surpassed many grim milestones as more than 3 million people died from COVID-19—a disease we barely knew existed when 2020 began. Our country's moral foundations were shaken, again and again, by horrifying violence and renewed calls for racial justice. And in spring of 2021, many individuals remained out of work and have been denied opportunities. In higher education alone, more than 650,000 jobs—one in eight—were lost.

These problems did not stop at the edge of Vanderbilt's campus. Rather, they invaded many aspects of our lives and presented one of the most challenging times our university—indeed, all of higher education—has ever faced.

There is a certain irony that I have spent a good portion of my academic career studying how leaders thrive in crisis. That research has included analyzing how leaders make effective decisions in a charged atmosphere of uncertainty, anxiety, and tight time constraints. My work has also explored the ways in which organizations weathering tough times can emerge stronger and positioned to flourish once the immediate dangers have passed.

Yet, just as many seasoned leaders experienced in the 2008–09 financial crisis, there is no substitute for leading an organization through once-in-a-lifetime events. Despite my background in research and years of experience as an executive, little could have prepared me (or anyone) for the multiple layers of ramifications that the past year's difficult circumstances have wrought.

I am deeply proud of what our leadership team, and the entire Vanderbilt community, has done amid so much adversity to ensure that we carry on with our nearly 150-year-old mission of research, education, and impact. While I don't claim that Vanderbilt had all the answers or did everything perfectly, we are proud of what was accomplished. We ensured that our students, student-athletes, researchers, faculty members, and staff could carry on with their work in as normal a way as possible. We were able to host students on campus for in-person learning—and reopen labs and essential offices—starting as early as the late spring of 2020. We pivoted and refined our operations as circumstances demanded, and we learned more about the pandemic and how to manage it. We engaged students, parents, alumni, and prospective students in new and innovative ways. Through it all, we were able to keep COVID-19 infections low. Our scientists and public health experts played a pivotal role in developing solutions to the pandemic, and our partnership with Vanderbilt University Medical Center proved invaluable.

Along the way, our collaborative approach took on new forms. We pulled together cross-functional teams in short order, further obliterating any of the silos that could make significant challenges more difficult to navigate. The traditional boundaries that are often associated with institutions of higher education were already low at Vanderbilt, but they continued to dissolve throughout this time, and the result will serve as a model for decades to come.

Operational and financial challenges were not the only hurdles we had to deal with. We also engaged in difficult conversations around race, diversity, inclusion, and partisanship—not just within the Vanderbilt community but across the nation.

In a spirit of expanding our knowledge, I offer the following observations about how we made decisions as an organization, built trust within our community, and bolstered Vanderbilt's reputation as an institution marked by its willingness to confront difficult circumstances with hard work, compassion, and commitment.

First Principles

How we frame a decision often foreshadows its outcome. When many universities were deciding whether to bring students back, their decisions were often framed around the question of whether or not to return to in-person learning. When that is the guiding framework, the less-demanding choice is to go remote. Though it may have worse long-term consequences, this path is easier and more expedient in the short term.

Our framing was different. We started from our first principles: Who are we as a university, and what is our purpose? The answers to those questions map directly back to our founding mission: we are a research university, and we are a residential college. That meant we needed our labs up and running as quickly as possible, and we wanted to invite students back onto campus and into our classrooms to the fullest extent possible.

That mindset extended into every corner of university life. For example, many of our peer institutions treated athletics as a dispensable activity. We didn't. For our student-athletes, being able to play their sports is one of the most important things in their lives. We share their commitment to college athletics at the highest level, so we did everything we could to make that possible for them. We, as a leadership team, did not want to tell them that they and their dreams—what they came to college to achieve—don't matter. Instead, we asked ourselves, "How can we make this work?" The same was true at the Blair School of Music. Choral singing and playing woodwind and brass instruments are particularly risky activities in the COVID-19 environment. Rather than abandon our commitment to those students' education and growth, we asked, "How can we make instruction possible for them?"

Those were always the questions. We did not see any particular activities as more or less important. Universities, at their very core, are focused on helping every single person in the community reach their full human potential. That goes for mathematicians and scientists as much as it does for performers, artists, and student-athletes. Our duty was always to continue to enable those pursuits as much as we could in the context of the pandemic.

That focus on our mission, in turn, motivated our people to go the extra mile. Our students, our faculty, our staff, our administration—we all were driven by a sense of shared purpose.

Know When to Act, Know When to Wait

Decision-making under extreme uncertainty is difficult. It's psychologically difficult, and it's procedurally difficult. For effective leaders, it is important to be able to endure prolonged, deep uncertainty—no matter how taxing. COVID-19 was an extreme case of profound and persistent uncertainty. Public health conditions shifted rapidly, often within days, and approaches that initially were believed to be true turned out to be false, and vice versa. And all too often we had no answers at all. For example, in the early days of the pandemic, people were encouraged not to wear masks. A few weeks later, masks became mandatory. Now the recommendation is to wear two masks.

To address these unique challenges, we were explicitly intentional in making our decisions when we had to—and not any earlier. New information or circumstances are always emerging, and they may lead to better outcomes than if you decide based on what you know now. In other words, we may know more about the world, or the world itself may change and provide better alternatives, if we wait.

That waiting was extremely difficult for people; they wanted to know whether we would be open in the fall. It wasn't just students, researchers, and parents who were eager to know. We were keenly aware that the decision had a major impact on our faculty, our staff, and many others. We eventually had to act so that people could make their own plans and their own decisions, but we needed to wait so that we could learn as much as possible. In fact, we decided quite late.

We could not be sure whether our plan would work, but we could reduce the risk. What I told our leadership team was that we wanted to be 80 percent confident we could succeed in reopening our campus, and that once we reached that level of confidence, we'd announce our decision. We would never get to 100 percent; we knew things would change.

The Need for Candid Dialogue

My favorite leadership maxim is credited to Cyrus the Great: "Diversity in council, unity in command." Perhaps the most important difficulty in high-stakes decision-making is the need to put all the information on the table. Yet decades of research have shown that traditionally in meetings, participants focus on the shared information and are reluctant to offer new perspectives. Good decision-making processes are conscious of these biases and try to overcome them. Once we developed our plans, we held a pre-mortem to anticipate what could go wrong. We also deployed "red teams" to punch holes in our plans. These approaches were invaluable; they uncovered hidden assumptions and changed circumstances, and they led to material improvements in our plans and protocols.

These techniques require a high level of trust among the management team. Here, as in so many other instances, our culture of collaboration and cooperation was essential. People were willing to have candid discussions in our atmosphere of mutual respect.

We're now applying this exercise to every major decision we make. The team that assembles the original plan must meet with a team that was not part of the decision-making process. That second team's job is to identify flaws in the plan, to stress-test the thinking that went into decisions, and to identify problems. We changed our entire fall academic calendar based on that exercise.

Here is where the value of time came into play: While we were making our plans for the fall 2020 semester, the assumptions on which we built our original solution became invalid. We had first wanted to start school as late as possible, but we later discovered it would be better to instead end the semester by Thanksgiving. That required us moving our target start date closer by two weeks. (The wave of cases that was anticipated toward the end of the summer did not materialize, and the concern then became that there would be a much larger wave in November—when people would travel and see family. And that is exactly what happened.) It was that "red team" exercise that led to action and, ultimately, to better decision-making.

We are proud of our success. But when you think about good decision-making, you can't evaluate it based on the outcome. You have to evaluate it in the context of the time when you made the decision. We know a lot of things now that we didn't know at that time. Things could have turned out much worse, but good decisions are still good decisions if the process is right—even if your luck is bad. All we could do was make the best decisions possible in the moment with the available information.

The Trust Radar

We knew from the earliest days, in March 2020, that as we were having to make difficult and complex decisions, we were doing so amid deep fears and widespread anxieties among our internal and external constituencies. The fact that peoples' health, their financial security, and their sense of justice were all under threat at the same time only heightened their need for direction and information. Similarly, our leadership team felt the heavy burdens of ensuring the health and safety of our community, of safeguarding our resources from sudden economic jolts, and of strengthening a sense of belonging within our university culture.

In the midst of crises like this, there is disorienting uncertainty and a deluge of information, some of it vital and some of it mere noise. It is imperative that leaders work quickly to build a sense of trust, which will gain them the much-needed space to properly evaluate decisions and the goodwill to be able to make the tough calls.

This may seem obvious, but in the heat of the moment, leaders often focus so much on managing day-to-day activities that they sometimes struggle to build and maintain that trust. Just as it is important to pursue an institution's long-term goals and priorities in the face of imminent decisions and tasks, leaders must also establish an enduring reputation for being reliable.

Research has informed what I call the "Trust Radar," comprised of four key factors that influence the level of trust that leaders are able to build in times of crisis.

The first factor is transparency.

This means different things to different stakeholders. What we shared with students and parents during the Return to Campus planning process was different from what we shared with faculty members—and this was because their concerns were different. We had to anticipate the information that would be most salient to these groups and then communicate it in ways that resonated most deeply with them—whether that was through a virtual town hall with a Q&A session, or through specific actions like implementing plexiglass shields for classroom safety.

Transparency, I have found, also hinges on clarity. Legalese or technical language is not as conducive to transparent communication as, say, a heartfelt speech that does not withhold information or hide behind complexities.

The second factor is expertise. A perceived lack of expertise can undermine trust quickly, and I cannot overstate the importance during this pandemic of having some of the world's top infectious disease researchers, public health experts, and clinicians from Vanderbilt University Medical Center deeply involved from the very beginning. Groups ranging from students and parents to faculty and staff, as well as our Board of Trust and alumni base, felt reassured about our actions—often because of the involvement of VUMC experts.

The third element is commitment. Because information changes so quickly in a crisis situation, and it's hard to establish even basic facts, it becomes vital for leaders to demonstrate their commitment to staying on top of the situation. The most powerful way to do this is for leaders to be highly visible and accessible, while demonstrating that they have taken charge. An example of this at Vanderbilt was the series of town halls we conducted through the summer. We didn't have all the answers, but it was important for us to show up and share what we did know. In addition, we consolidated information onto a single COVID-19 website that was the source of truth for all our public decisions. The town halls and the website were complemented by a steady stream of direct email and video communication developed for multiple audiences.

Finally, and I believe most importantly, leaders must show empathy. This is perhaps the easiest thing to do, but it is usually one of the most overlooked components of building trust. We have all been affected by COVID-19. Maybe one has felt isolated and alone, has lost their job, or has had a friend or family member get sick or, worse, die. Even just witnessing the year's multiple national traumas—not to mention Nashville's own share of suffering—could be devastating. These are deep, painful emotions that greatly affect all stakeholders. Not only was it important for our leadership team to convey a sense of empathy, but we also needed to back it up with special attention to harnessing resources, like mental health and well-being services and financial hardship funds, to let our university family know that we cared about them and their loved ones.

Foster and Embrace Community

In the context of crises like the pandemic—as well as troubling events like the Nashville tornadoes, the Christmas Day bombing, and the national social unrest—it is vital that an institution fully realizes the importance of action as one community. What I came to call the One Vanderbilt spirit of compassion and caring that was demonstrated through the past year helped to counteract people's fear and uncertainty. We paid special attention to ensuring that the needs of our community were recognized and addressed to the extent we could. At the very least, we did not want our actions to be perceived as cold or merely transactional.

In research I conducted with the Institute for Management Development's Jennifer Jordan and Columbia University's Adam D. Galinsky, we called this approach the Good Samaritan

Principle: caring combined with competence. This concept includes the following elements:

- Acting with authentic concern by serving the broader needs of the community rather than pursuing our short-term interests
- Acting with competence and confidence
- Communicating in a way that is helpful to multiple stakeholders and is not self-serving

Depending on how organizations perform in these critical moments of crisis, and on how their actions are perceived by the community, these periods can create turning points that lead to a positive and lasting impact.

Our Proudest Moment

I have talked often about this being our proudest moment. Part of that idea stemmed from a Vanderbilt trustee suggesting that I read Erik Larson's recently published *The Splendid and the Vile*, a history of UK prime minister Winston Churchill during the bombing of London. In the midst of unfathomable danger, he embraced the challenge and knew that it would test the character and mettle of the British nation. Churchill did not shy away from explaining the harsh reality and many setbacks to the British public, yet he remained ever the optimist—certain that his country would find its way to a brighter day. Throughout the past year, I was often buoyed by thoughts of Churchill's spirit and of his ability to be both utterly realistic and full of hope.

How we act during crises shapes how we are remembered. The actions we take, and our reasons for taking them, have a lasting impact on our personal legacies and on the legacy of Vanderbilt.

During these most difficult of times, our entire community—faculty, staff, students, parents, alumni—aligned around a shared purpose grounded in a common set of values, and together overcame the biggest challenges ever experienced in the history of the university. It was this spirit of unity that made this a year like no other.

ACKNOWLEDGMENTS

When Chancellor Daniel Diermeier first suggested the idea of assembling a book capturing Vanderbilt's experience navigating the challenges of the past year, many of us who would get involved with this project nodded in agreement as we silently wondered if it would ever become a reality. For Chancellor Diermeier, there was never any question. He knew it would require a great deal of work done in a short amount of time. He also knew that it would have to be completed during a period when many people were already working their hardest. Yet, importantly, he took the long view required of anyone leading an enduring and historically important institution such as Vanderbilt. As with the 1918 Flu Pandemic, the Great Depression, both world wars, Vietnam, or the civil rights era, it was critical to record Vanderbilt's place in this indelible history—and to do so in real time, not by examining this experience through the distorting lens of hindsight years from now. For that motivation, that wisdom, and that foresight, we are all grateful. We are also thankful for Chancellor Diermeier's time and his office's sustained support of this project.

Thank you also to Susan R. Wente, Vanderbilt provost, interim chancellor, and now Wake Forest University's fourteenth president, the first woman to lead both institutions.

The university community will always be profoundly grateful for her determined and compassionate leadership under the most difficult circumstances. For this book, she devoted hours of discussion and review, as well as critical behind-the-scenes insights and observations.

Many thanks, as well, to members of Vanderbilt's leadership team: Dr. André Churchwell, vice chancellor for equity, diversity, and inclusion and chief diversity officer; Steve Ertel, vice chancellor for communications; Eric Kopstain, vice chancellor for administration; Candice Lee, vice chancellor for athletics and university affairs and athletic director; John Lutz, vice chancellor for information technology and interim vice chancellor for development and alumni relations; Ruby Shellaway, vice chancellor, general counsel, and university secretary; Brett Sweet, vice chancellor for finance and chief financial officer; and Dawn Turton, chief of staff to Chancellor Diermeier. All of you somehow carved out time in your schedules to help with this book, despite working seven days a week since March 2020 with little or no breaks.

From the Vanderbilt Board of Trust, our team is indebted to chairman Bruce Evans, BE'81, for his help in reconstructing events, offering personal perspectives, tracking down correspondence, and reviewing manuscripts. In addition, Maribeth Geracioti and Lissa Allen proved extraordinarily helpful in locating key information.

Melanie Moran, associate vice chancellor for university relations, and Rachel Albright, strategic consultant for the Ingram Group, have a superhuman capacity to make things happen, solve problems, develop strategies, and manage head-spinning levels of chaos and complexity. They were key to making this book happen. Full stop.

Words can hardly do justice to the writing brilliance, the Olympian capacity for hard work, the graceful patience, and the countless hours of interviews, research, drafting, and editing by Kevin Jones, a one-of-a-kind writer based in Seattle, and Lucie Alig, an executive writer in Vanderbilt's division of communications. This book is as much yours as anyone's. Thank you. Thank you. Thank you.

Finally, to the Vanderbilt community and my fellow Commodores, you have been a source of awe and inspiration for me for nearly thirty years since I first began wandering the paths of campus as a wide-eyed undergraduate. That has been especially true as I've heard from friends, former classmates, current students, professors, parents, fellow staff members, coaches, and others during the past year. Your resilience, your optimism, and the keen intelligence you bring to just about every situation make our university proud.

As Chancellor Diermeier likes to say, this indeed is our finest hour.

—Ryan Underwood

ENDNOTES

1 "JFK delivers iconic 'We choose to go to the Moon' speech," This Day in History, Space Center Houston, September 9, 2019, https://spacecenter.org/this-day-in-history-jfk-delivers-iconic-we-choose-to-go-to-the-moon-speech/; "John F. Kennedy Moon Speech - Rice Stadium," NASA, https://er.jsc.nasa.gov/seh/ricetalk.htm.

2 Derrick Bryson Taylor, "A Timeline of the Coronavirus Pandemic," *New York Times*, March 17, 2021, https://www.nytimes.com/article/coronavirus-timeline.html.

3 Taylor, "A Timeline of the Coronavirus Pandemic."

4 "University Monitoring Coronavirus Outbreak," Vanderbilt News, Vanderbilt University, January 29, 2020, https://news.vanderbilt.edu/2020/01/29/university-monitoring-coronavirus-outbreak/.

5 "Jan 26 - Feb 1, 2020," Events@Vanderbilt, Vanderbilt University, https://events.vanderbilt.edu/week/date/20200126.

6 Information from Chancellor's Lecture Series Archives, Vanderbilt University, https://www.vanderbilt.edu/chancellor/cls-archives/.

7 Taylor, "A Timeline of the Coronavirus Pandemic."

8 "Timeline: How the Global Coronavirus Pandemic Unfolded," Reuters.com, June 28, 2020, https://www.reuters.com/article/us-health-coronavirus-timeline/timeline-how-the-global-coronavirus-pandemic-unfolded-idUSKBN23Z0UW.

9 "Timeline," Reuters, June 28, 2020.

10 Taylor, "A Timeline of the Coronavirus Pandemic."

11 Jun Zheng, "SARS-CoV-2: An Emerging Coronavirus That Causes

a Global Threat," *International Journal of Biological Sciences* 16, no. 10 (March 15, 2020): 1678-85, https://www.ncbi.nlm.nih.gov/pmc/articles/PMC7098030/.

12 "Timeline," Reuters.com, June 28, 2020.

13 Susan Wente PowerPoint Presentation to the Vanderbilt Board of Trust, "Feb BOT COVID Research Contributions draft_012521," Slide 6.

14 Morgan Kroll, "Coronavirus Expert Mark Denison Shares COVID-19 Research and Insights," Vanderbilt News, Vanderbilt University, December 16, 2020, https://news.vanderbilt.edu/2020/12/16/coronavirus-expert-mark-denison-shares-covid-19-research-and-insights/.

15 Marissa Shapiro, "Vanderbilt Researchers Take Leadership Role in COVID-19 Vaccine Development," Vanderbilt News, Vanderbilt University, December 18, 2020, https://news.vanderbilt.edu/2020/12/18/vanderbilt-researchers-take-leadership-role-in-covid-19-vaccine-development/.

16 Michael Blanding, "Shot in the Arm," *Vanderbilt Magazine*, Spring 2021, p. 32.

17 Bill Snyder and Kathy Whitney, "The Front Lines: Vanderbilt Physicians, Researchers Join Worldwide Fight Against COVID-19," *Vanderbilt Magazine*, Spring 2020, p. 19.

18 "Gilead Sciences Initiates Two Phase 3 Studies of Investigational Antiviral Remdesivir for the Treatment of COVID-19," Gilead Sciences (press release), February 26, 2020, https://www.gilead.com/news-and-press/press-room/press-releases/2020/2/gilead-sciences-initiates-two-phase-3-studies-of-investigational-antiviral-remdesivir-for-the-treatment-of-covid-19.

19 Snyder and Whitney, "The Front Lines," 19.

20 Snyder and Whitney, "The Front Lines," 19.

21 Snyder and Whitney, "The Front Lines," 19.

22 Spencer Turney, "Breathing Easier: Interdisciplinary Team Develops Open-source Ventilator Design," *Vanderbilt Magazine*, Spring 2020, p. 13.

23 Liz Entman, "Visual Aid," *Vanderbilt Magazine*, Spring 2020, p. 15.

24 *Vanderbilt Magazine*, Spring 2020, p. 4.

25 Jessica Bliss, "Tennessee Tornadoes' Path of Terror," *Tennessean*, March 25, 2020, https://www.tennessean.com/pages/interactives/news/graphics/march-2020-tornado-path-nashville-cookeville-putnam-tennessee/.

26 "University Update on March 3 Tornado Impact," *Vanderbilt News*, Vanderbilt University, March 3, 2020, https://news.vanderbilt.edu/2020/03/03/statement-on-march-3-tornadoes/.

27 [[author]]"Picking Up the Pieces," *Vanderbilt Magazine*, Spring 2020, p. 2.

28 Karen Weintraub, "From a Biogen Conference to a Homeless Shelter: Researchers Track Coronavirus Infections from 'Super-spreader' Events," *USA TODAY*, August 26, 2020, https://www.usatoday.com/story/news/2020/08/26/how-superspreader-events-biogen-conference-incubated-coronavirus-research/3437458001/.

29 Eva Durchholz, Rachel Friedman, and Immanual John Milton, "Vanderbilt Student Tests Positive for Coronavirus in Chicago," *Vanderbilt Hustler*, March 5, 2020, https://vanderbilthustler.com/30926/featured/breaking-vanderbilt-student-tests-positive-for-coronavirus-in-chicago/.

30 "Message from Interim Chancellor and Provost Wente Regarding COVID-19," MyVU, Vanderbilt University, March 5, 2020, https://news.vanderbilt.edu/2020/03/05/a-message-from-interim-chancellor-and-provost-wente-regarding-covid-19/.

31 Susan Svrluga, "Stanford, Others Switch to Online Classes Temporarily Amid Coronavirus Fears," *Washington Post*, March 8, 2020, https://www.washingtonpost.com/education/2020/03/08/stanford-cancels-in-person-classes-temporarily-amid-coronavirus-fears/.

32 "Mar. 7, 2020 – Message to Parents Regarding Students' Return from Spring Break," Return to Campus, Vanderbilt University, March 7, 2020, https://www.vanderbilt.edu/coronavirus/2020/03/07/mar-7-2020-dean-of-students-message-about-returning-from-spring-break/.

33 "Mar. 7, 2020 – Vice Provost for Faculty Affairs Message about Continuity Planning for Teaching," Return to Campus, Van-

derbilt University, March 7, 2020, https://www.vanderbilt.edu/coronavirus/2020/03/07/mar-7-2020-vice-provost-for-faculty-affairs-message-about-continuity-planning-for-teaching/.

34 Mariah Timms, "Coronavirus in Tennessee: Timeline of Cases, Closures and Changes as COVID-19 Moves In," *Tennessean*, March 25, 2020, https://www.tennessean.com/story/news/health/2020/03/26/coronavirus-tennessee-timeline-cases-closures-and-changes/2863082001/.

35 Mark Bandas, email to Vanderbilt University student body, March 8, 2020, https://www.vanderbilt.edu/email-creator/2020/03/coronavirus-update-keeping-our-community-safe/.

36 Anna Irrera and Steve Stecklow, "How Elite U.S. College Students Brought COVID-19 Home from Campus," Reuters, April 2, 2020, https://www.reuters.com/article/us-health-coronavirus-usa-vanderbilt-exc/exclusive-how-elite-u-s-college-students-brought-covid-19-home-from-campus-idUSKBN21K2CJ.

37 "Mar. 11, 2020 – Message to Community about Alternative Education for Remainder of Semester," Return to Campus, Vanderbilt University, March 11, 2020, https://www.vanderbilt.edu/coronavirus/2020/03/11/community-alternative-education/.

38 Irrera and Stecklow, "How Elite U.S. College Students Brought COVID-19 Home from Campus."

39 Brett Kelman, "Vanderbilt University Undergrads Must Leave Campus, Spring Semester Classes Moving Online," *Tennessean*, March 11, 2020, https://www.tennessean.com/story/news/health/2020/03/11/vanderbilt-university-moving-spring-classes-online-undergrads-must-leave-campus-coronavirus-spread/5026434002/.

40 "Move-Out Information," Housing and Residential Experience, Vanderbilt University, https://www.vanderbilt.edu/ohare/move-out-instructions-and-information/#item4065.

41 Evan Monk, "Pandemic Profiles: International Student Safa Shahzad," *Vanderbilt Hustler*, December 29, 2020, https://vanderbilthustler.com/37336/featured/pandemic-profiles-international-student-safa-shahzad/.

42 "Board of Trust Establishes Ad Hoc Committee to Support University's Coronavirus Response," myVU, Vanderbilt University, March 17, 2020, https://news.vanderbilt.edu/2020/03/17/board-of-trust-establishes-ad-hoc-committee-to-support-universitys-coronavirus-response/.
43 Craig Stephenson, "SEC Cancels All Spring Sports Competition, Including Baseball Tournament, Spring Football Games," AL.com, March 17, 2020, https://www.al.com/sports/2020/03/sec-cancels-all-spring-sports-competition-including-baseball-tournament.html.
44 Andrew Smalley, "Higher Education Responses to Coronavirus (COVID-19)," National Conference of State Legislators, December 28, 2020, https://www.ncsl.org/research/education/higher-education-responses-to-coronavirus-covid-19.aspx.
45 Shapiro, "Vanderbilt Researchers Take Leadership Role in COVID-19 Vaccine Development."
46 "WATCH: A Word with Wente—A Discussion with Dr. Jeff Balser," myVU, Vanderbilt University, June 22, 2020, https://news.vanderbilt.edu/2020/06/22/watch-a-word-with-wente-a-discussion-with-dr-jeff-balser/.
47 Timms, "Coronavirus in Tennessee."
48 Staff reports, "Leading Authorities," *Vanderbilt Magazine*, Spring 2020, p. 5.
49 "2020 Commencement Postponed a Year Due to COVID-19 Outbreak," Vanderbilt News, Vanderbilt University, March 25, 2020, https://news.vanderbilt.edu/2020/03/25/commencement2020-postponed/.
50 Amy Wolf, "Songwriting Seminar Inspires Personal Tributes to Class of 2020," myVU, Vanderbilt University, May 6, 2020, https://news.vanderbilt.edu/2020/05/06/songwriting-seminar-inspires-personal-tributes-to-class-of-2020/.
51 Susan R. Wente, "Open Mind: VU and VUMC—Stronger Together," myVU, Vanderbilt University, April 6, 2020, https://news.vanderbilt.edu/2020/04/06/open-mind-vu-and-vumc-stronger-together/.
52 "A Message from Interim Chancellor and Provost Susan R.

Wente," Faculty Senate, Vanderbilt University, https://www.vanderbilt.edu/facultysenate/assembly/2020-spring-assembly/index.php.

53 "Remote Learning Enablement at Vanderbilt University," Vanderbilt University YouTube channel, April 14, 2020, https://www.youtube.com/watch?v=IcfYxuJjAdk.

54 "Remote Learning," YouTube video, April 14, 2020.

55 Scott Gottlieb et al., "National Coronavirus Response: A Road Map to Reopening," American Enterprise Institute, March 29, 2020, https://www.aei.org/research-products/report/national-coronavirus-response-a-road-map-to-reopening/.

56 Chris Francescani, "Timeline: The First 100 Days of New York Gov. Andrew Cuomo's COVID-19 Response," abcnews.go.com, June 17, 2020, https://abcnews.go.com/US/News/timeline-100-days-york-gov-andrew-cuomos-covid/story?id=71292880.

57 Francescani, "Timeline."

58 Return-to-Campus-Virtual-Town-Hall-2020-5-13-myVU-1-3.pdf.

59 "Apr. 17, 2020 – A Message from Interim Chancellor Susan R. Wente: The Economic Impact of COVID-19," Return to Campus, Vanderbilt University, April 17, 2020, https://www.vanderbilt.edu/coronavirus/2020/04/17/apr-17-2020-a-message-from-interim-chancellor-susan-r-wente-the-economic-impact-of-covid-19/.

60 "About the endowment," Giving to Vanderbilt, Vanderbilt University, https://giving.vanderbilt.edu/endowment/about.php.

61 "Helping Hand: VU to Use $2.8 Million in Federal Funding to Support Students," *Vanderbilt Magazine*, Spring 2020, p. 6.

62 "May 7, 2020 – Next Steps for Vanderbilt Campus as Nashville Enters Phase 1 of Reopening," Return to Campus, Vanderbilt University, May 7, 2020, https://www.vanderbilt.edu/coronavirus/2020/05/07/may-7-2020-next-steps-for-vanderbilt-campus-as-nashville-enters-phase-1-of-reopening/.

63 Ann Marie Deer Owens, "Wente, Kopstain Discuss Return to Campus Plan at Virtual Town Hall," Vanderbilt News, Vanderbilt University, May 15, 2020, https://news.vanderbilt.edu/2020/05/15/wente-kopstain-discuss-return-to-campus-

plan-at-virtual-town-hall/.

64 Vanderbilt University Virtual Town Hall presentation, May 13, 2020.

65 Town Hall presentation, May 13, 2020.

66 Town Hall presentation, May 13, 2020.

67 Lilah Burke, "Certainly Uncertain," *Inside Higher Education*, April 30, 2020, https://www.insidehighered.com/news/2020/04/30/what-does-intent-reopen-mean.

68 "Return to Campus Phases," Return to Campus, Vanderbilt University, https://www.vanderbilt.edu/coronavirus/return-to-campus-phases/#phase-one.

69 Owens, "Wente, Kopstain Discuss Return to Campus Plan at Virtual Town Hall."

70 2020-4-16 Full BOT Meeting Materials.pdf, p. 208.

71 2020-4-16 Full BOT Meeting Materials.pdf, p. 36.

72 2020-4-16 Full BOT Meeting Materials.pdf, p. 35.

73 2020-4-16 Full BOT Meeting Materials.pdf, p. 38.

74 Jenna Somers, "Engineering Lab Returns During Vanderbilt's Research Ramp-up to Advance Research In Neurodegeneration," Vanderbilt News, Vanderbilt University, November 16, 2020, https://engineering.vanderbilt.edu/news/2020/engineering-lab-returns-during-research-ramp-up-to-advance-research-in-neurodegeneration/.

75 Liz Entman, "Vanderbilt University Screening Tool Assesses COVID-19 Risk," Research News@Vanderbilt, Vanderbilt University, June 1, 2020, https://news.vanderbilt.edu/2020/06/01/vanderbilt-university-screening-tool-assesses-covid-19-risk/.

76 Excerpted from "Shot in the Arm," by Michael Blanding, https://vanderbilt.app.box.com/s/ttho bmvlo875movqbcoskmtx3fixz9ex.

77 Snyder and Whitney, "The Front Lines," 20.

78 Snyder and Whitney, "The Front Lines," 20.

79 "New Trial Plans Rigorous Test of Convalescent Plasma Therapy," Discover, Vanderbilt University Medical Center, May 20, 2020, https://discover.vumc.org/2020/05/trial-explores-nuances-of-convalescent-plasma-for-covid-19/.

80 "Critical Incident Message—Statement on the death of George

Floyd," Equity, Diversity and Inclusion, Vanderbilt University, May 29, 2020, https://www.vanderbilt.edu/diversity/2020/05/29/george-floyd/.

81 "Critical Incident Message—Statement on the Death of George Floyd."

82 "Vanderbilt Statement on Racial Injustice in Our Society," Vanderbilt News, Vanderbilt University, May 31, 2020, https://news.vanderbilt.edu/2020/05/31/wente-0531-racial-injustice/.

83 "Virtual Gathering for Faculty with Incoming Chancellor Daniel Diermeier," Vanderbilt University YouTube channel, June 9, 2020, https://www.youtube.com/watch?t=180&v=0_ti_ksWgKc&feature=youtu.be.

84 "Virtual Gathering for Students and Families," Vanderbilt University YouTube channel, June 3, 2020, https://www.youtube.com/watch?v=gGLd1LSjp0A&feature=youtu.be.

85 "Memorial Service Against Racism," Equity, Diversity and Inclusion, Vanderbilt University, June 29, 2020, https://www.vanderbilt.edu/diversity/memorial-service/.

86 Jalen Blue, "Hundreds Gather for Virtual Memorial Service Against Racisms Hosted by Churchwell," Vanderbilt News, Vanderbilt University, July 6, 2020, https://news.vanderbilt.edu/2020/07/06/hundreds-gather-for-virtual-memorial-service-against-racism-hosted-by-churchwell/.

87 "Our Commitment to an Inclusive Vanderbilt," myVU, Vanderbilt University, July 1, 2020, https://news.vanderbilt.edu/2020/07/01/our-commitment-to-an-inclusive-vanderbilt/.

88 Immanual John Milton, "Video surfaces of Vanderbilt student using racial slur, ties to Greek life," *Vanderbilt Hustler*, July 20, 2020, https://vanderbilthustler.com/33256/featured/video-surfaces-of-vanderbilt-student-using-racial-slur-ties-to-greek-life/.

89 "Vanderbilt University Statements on Greek Life," Vanderbilt News, Vanderbilt University, July 7, 2020, https://news.vanderbilt.edu/2020/07/07/vanderbilt-university-statement-on-greek-life/.

90 Taylor, "A Timeline of the Coronavirus Pandemic."

91 "Timeline," Reuters.com, June 28, 2020.
92 Taylor, "A Timeline of the Coronavirus Pandemic."
93 Janice Hopkins Tanne, "Covid-19 Cases Increase Steeply in US in South and West," The BMJ, *British Medical Journal* 2020; 369: m2616, June 29, 2020.
94 Davidson County, Tennessee, COVID-19 dashboard, https://nashville.maps.arcgis.com/apps/MapSeries/index.html?appid=30dd8aa876164e05ad6c0a1726fc77a4.
95 Jon Garcia and Natalie Neysa Alund, "What to Know About Phase Two in the Plan to Reopen Nashville," *Tennessean,* May 18, 2020, https://www.tennessean.com/story/news/local/2020/05/18/nashville-reopening-plan-phase-2/5213270002/.
96 "June 16, 2020 – Our Plan for the Fall Semester," Return to Campus, Vanderbilt University, June 16, 2020, https://www.vanderbilt.edu/coronavirus/2020/06/16/june-16-2020-our-plan-for-the-fall-semester/.
97 "Fall Semester," June 16, 2020.
98 Robert Kelchen, "This Will Be One of the Worst Months in the History of Higher Education," *The Chronicle of Higher Education*, July 7, 2020, https://www.chronicle.com/article/this-will-be-one-of-the-worst-months-in-the-history-of-higher-education.
99 Susan Dynarski, "College Is Worth It, but Campus Isn't," *New York Times,* June 29, 2020, updated July 3, 2020, https://www.nytimes.com/2020/06/29/business/college-campus-coronavirus-danger.html?searchResultPosition=1.
100 Janet H. Murray, "O-Rings, Groupthink and Campus Reopenings," *Inside Higher Ed* August 17, 2020, https://www.insidehighered.com/views/2020/08/17/college-officials-bringing-students-back-campus-are-challenger-engineers-building.
101 Rachel Friedman, "Peer Institutions Take Divergent Approaches to Bringing Students Back to Campus this Fall," *Vanderbilt Hustler*, August 16, 2020, https://vanderbilthustler.com/33642/featured/peer-institutions-take-divergent-approaches-to-bringing-students-back-to-campus-this-fall/.
102 Vanderbilt AAUP Guest Writer, "Risky for No Reason? Faculty Demand Answers," *Vanderbilt Hustler*, August 21, 2020,

https://vanderbilthustler.com/33751/featured/guest-editorial-risky-for-no-reason-faculty-demand-answers/.

103 The Parent Plan – Vanderbilt website, https://www.theparent-plan-vanderbilt.net/.

104 Lindsay Ellis, "Colleges Hoped for an In-Person Fall. Now the Dream Is Crumbling," *The Chronicle of Higher Education*, July 20, 2020, https://www.chronicle.com/article/Colleges-Hoped-for-an/249206?cid2=gen_login_refresh&cid=gen_sign_in.

105 Jalen Blue, "University Launches Public Health Advisory Task Force in Response to COVID-19," myVU, Vanderbilt University, March 23, 2020, https://news.vanderbilt.edu/2020/03/23/university-launches-public-health-advisory-task-force-in-response-to-covid-19/#new_tab.

106 "Andrea George," Division of Administration, Vanderbilt University, https://www.vanderbilt.edu/administration/semo/meet-thestaff.php.

107 "August 10, 2020 – Return to Campus Update," Return to Campus, Vanderbilt University, https://www.vanderbilt.edu/coronavirus/2020/08/10/august-10-2020-return-to-campus-update/.

108 "Watch: Town Hall for Faculty Discusses Classroom Safety," myVU, Vanderbilt University, July 22, 2020, https://news.vanderbilt.edu/2020/07/22/watch-town-hall-for-faculty-discusses-classroom-safety/.

109 "Quick Facts," Vanderbilt University, https://www.vanderbilt.edu/about/facts/.

110 Zoe Abel, "First-years Living in Carmichael Towers Struggle to Find Community Outside Commons," *Vanderbilt Hustler*, September 9, 2020, https://vanderbilthustler.com/34181/featured/first-years-living-in-carmichael-towers-struggle-to-find-community-outside-commons/.

111 Avery Muir, "Widespread Zoom Outages Cause Faculty and Students to Adapt First-Day-of-Class Plans," *Vanderbilt Hustler*, August 24, 2020, https://vanderbilthustler.com/33824/featured/widespread-zoom-outages-cause-faculty-and-students-to-adapt-first-day-of-class-plans/.

112 "August 12, 2020—A Message to Faculty on Fulfilling Our Mis-

sion as We Return to Campus," Return to Campus, Vanderbilt University, https://www.vanderbilt.edu/coronavirus/2020/08/12/aug-12-2020-a-message-to-faculty-on-fulfilling-our-mission-as-we-return-to-campus/.

113 Gentry Estes, "Vanderbilt Didn't Look So Eager, but Turns Out, It Wants a Football Season, Too," *Tennessean*, June 16, 2020, https://www.tennessean.com/story/sports/college/vanderbilt/2020/06/17/vanderbilt-didnt-look-so-eager-but-wants-football-season-too/3199084001/.

114 Number of students living on campus as of September 2020, according to the Vanderbilt University Department of Planning and Institutional Effectiveness.

115 Hoon Kim, "Incoming First-years Scramble in Reaction to Vanderbilt's Return to Campus Plan," *Vanderbilt Hustler*, June 18, 2020, https://vanderbilthustler.com/32880/featured/incoming-first-years-scramble-in-reaction-to-vanderbilts-return-to-campus-plan/.

116 Return to Campus website, https://www.vanderbilt.edu/coronavirus/community/resources/fall-2020-campus-life/

117 Instagram post, https://www.instagram.com/p/CEFMzXwJc2X/.

118 "A message from Chancellor Diermier to Students: Prove the Doubters Wrong," Vanderbilt University YouTube channel, August 24, 2020, https://www.youtube.com/watch?v=ixBciItJOcU.

119 Muir, "Widespread Zoom Outages."

120 Susan R. Wente email, August 27, 2020, https://t.e2ma.net/message/h97w7d/1kykj6, referenced in *Vanderbilt Hustler*, August 27, 2020, https://vanderbilthustler.com/33877/featured/mandatory-testing-to-begin-aug-31-for-all-undergraduates-authorized-to-be-on-campus/.

121 Simon Gibbs, "Vanderbilt Pauses Football Activities, Moves Some Student-athletes into Isolation Housing," *Vanderbilt Hustler*, August 21, 2020, https://vanderbilthustler.com/33759/featured/breaking-vanderbilt-pauses-football-activities-moves-some-student-athletes-into-isolation-housing/.

122 Homepage, Vault Health, https://www.vaulthealth.com/.

123 "Contract Tracing for COVID 19," U.S Centers for Disease

Control and Prevention website, https://www.cdc.gov/coronavirus/2019-ncov/php/contact-tracing/contact-tracing-plan/contact-tracing.html.

124 "About," REDCap, https://projectredcap.org/about/.

125 PowerPoint presentation to the Vanderbilt Board of Trust Ad Hoc Committee on University Response to COVID-19, "Item 2 Update on Health and Well-Being of the Campus," September 3, 2020, slides 6 and 7.

126 Thomas Hum, "University to investigate reported 50-200 person gathering at Commons," *Vanderbilt Hustler*, August 30, 2020, https://vanderbilthustler.com/33941/featured/university-to-investigate-reported-50-200-person-gathering-at-commons/.

127 Source for these examples: "Four Weeks in, Faculty are Successfully Navigating New Teaching Landscape," myVU, Vanderbilt University, September 23, 2020, https://news.vanderbilt.edu/2020/09/23/four-weeks-in-faculty-are-successfully-navigating-new-teaching-landscape/.

128 Chase Mandell quoted in "Masked and Answered," Instagram, February 26, 2021, https://www.instagram.com/p/CLxR-D_hLjk/?igshid=12g9sxbvgo810.

129 Macy Su quoted in "Masked and Answered," Instagram, February 20, 2021, https://www.instagram.com/p/CLfJEuSBmnQ/?igshid=1vm70cz0ttw7x.

130 PowerPoint presentation, "Item 2 Update on the Well Being of the Campus," Slide 13, and other sources

131 Abel, "First-years Living in Carmichael Towers."

132 PowerPoint presentation, "Item 2 Update on Health and Well-Being of the Campus," September 3, 2020, slide 14.

133 Simon Gibbs, "Multiple Vanderbilt Football and Soccer Players Test Positive for COVID-19," *Vanderbilt Hustler*, September 2, 2020, https://vanderbilthustler.com/34019/featured/multiple-vanderbilt-football-and-soccer-players-test-positive-for-covid-19/.

134 "COVID-19 Positive Cases Among On-Campus Vanderbilt Community Members," Return to Campus, Vanderbilt University, https://www.vanderbilt.edu/coronavirus/fall2020dashboard/.

135 "Testing Percent Positive 7-Day Average," Davidson County COVID-19 Dashboard, https://nashville.maps.arcgis.com/apps/MapSeries/index.html?appid=30dd8aa876164e05ad6c0a1726fc77a4.
136 Madeline King and Zoe Abel, "False Positive Vault Results Erroneously Send Five Students into Isolation, Others into Quarantine," *Vanderbilt Hustler*, October 4, 2020, https://vanderbilthustler.com/35020/featured/false-positive-vault-results-erroneously-send-five-students-into-isolation-others-into-quarantine/.
137 Lacy Paschal, "Oct. 7, 2020—Our Plan for the Spring Semester," Return to Campus, Vanderbilt University, October 7, 2020, https://www.vanderbilt.edu/coronavirus/2020/10/07/spring-2021-plans/.
138 "Testing Percent Positive 7-Day Average," Davidson County COVID-19 Dashboard, https://nashville.maps.arcgis.com/apps/MapSeries/index.html?appid=30dd8aa876164e05ad6c0a1726fc77a4.
139 "Oct. 30, 2020—Staying Strong as We Head into Halloween," Return to Campus, Vanderbilt University, https://www.vanderbilt.edu/coronavirus/2020/10/31/oct-30-2020-staying-strong-as-we-head-into-halloween/.
140 Video about the Vaughn Home: https://www.vanderbilt.edu/innervu/news/vaughn-house-haunted-tour.
141 "COVID-19 Positive Cases Among On-Campus Vanderbilt Community Members," Return to Campus, Vanderbilt University, https://www.vanderbilt.edu/coronavirus/fall2020dashboard/.
142 Shapiro, "Vanderbilt Researchers Take Leadership Role in COVID-19 Vaccine Development."
143 Maria Cramer, "Dolly Parton Donated $1 Million to Help Develop a Coronavirus Vaccine," *New York Times*, November 18, 2020, https://www.nytimes.com/2020/11/18/world/dolly-parton-donated-1-million-to-help-develop-a-coronavirus-vaccine.html.
144 Taylor, "A Timeline of the Coronavirus Pandemic."
145 Vanderbilt Hustler Editorial Board, "Thank you, Vanderbilt," *Vanderbilt Hustler*, January 25, 2021, https://vanderbilthustler.

com/37645/featured/staff-editorial-thank-you-vanderbilt/.
146 Women's basketball news release, November 17, 2020, https://vucommodores.com/matchup-with-tennessee-tech-canceled/.
147 Betsy Goodfriend, "A Timeline of How Vanderbilt Women's Basketball Roster Shrunk to Just Seven Healthy Players," *Vanderbilt Hustler*, January 23, 2021, https://vanderbilthustler.com/37595/featured/a-timeline-of-how-vanderbilt-womens-basketball-roster-shrunk-to-just-seven-healthy-players/.
148 Adam Sparks, "Vanderbilt Women's Basketball Player Demi Washington Out for Season with Myocarditis," *Tennessean*, December 7, 2020, https://www.tennessean.com/story/sports/college/vanderbilt/2020/12/07/vanderbilt-womens-basketball-player-covid-19-myocarditis-demi-washington-out-season/6484351002/.
149 Amy Wolf, "Students Surprise Professor with Special Act of Gratitude," myVU, Vanderbilt University, December 18, 2020, https://news.vanderbilt.edu/2020/12/18/students-surprise-professor-with-special-act-of-gratitude/.
150 Ryan Suddath, "Vanderbilt Fall vs. Spring: What's the Difference?," *Vanderbilt Hustler*, February 8, 2021, https://vanderbilthustler.com/38062/featured/vanderbilt-fall-vs-spring-whats-the-difference/.
151 Amy Wolf, "Jason Isbell, Entertainers and University Leaders Come Together to Thank Vanderbilt Community," Vanderbilt News, Vanderbilt University, December 18, 2020, https://news.vanderbilt.edu/2020/12/18/jason-isbell-entertainers-and-university-leaders-come-together-to-thank-vanderbilt-community/.
152 Ryan Suddath, "Vanderbilt University Medical Center Begins COVID-19 Vaccinations," *Vanderbilt Hustler*, December 18, 2020, https://vanderbilthustler.com/37226/featured/vanderbilt-university-medical-center-begins-covid-19-vaccinations/.
153 Data—and other campaign info throughout—came from pitch deck from Daniel: https://vanderbilt.app.box.com/file/771594009695.
154 Kurt Streeter, "For This College Athlete, Covid-19 Was Just the

Start of a Nightmare," *New York Times*, February 12, 2021, https://www.nytimes.com/2021/02/12/sports/ncaabasketball/college-sports-myocarditis.html.

155 Streeter, "For This College Athlete, Covid-19 Was Just the Start of a Nightmare."